D0041116

The
Fragile
Species

The Fragile Species

Lewis Thomas

A ROBERT STEWART BOOK

CHARLES SCRIBNER'S SONS · NEW YORK

Maxwell Macmillan Canada · *Toronto*
Maxwell Macmillan International
New York · Oxford · Singapore · Sydney

Charles Scribner's Sons
Macmillan Publishing Company
866 Third Avenue
New York, NY 10022

Maxwell Macmillan Canada, Inc.
200 Eglinton Avenue East
Suite 200
Don Mills, Ontario M3C 3N1

Macmillan Publishing Company is part of the
Maxwell Communication Group of Companies.

Library of Congress Cataloging-in-Publication Data

Thomas, Lewis, 1913–
  The fragile species / Lewis Thomas
    p. cm.
  "A Robert Stewart book."
  Includes index.
  ISBN 0-684-19420-1
  1. Medicine. 2. Biology I. Title.
  R708.T48 1992
  610′.1—dc20   91-34315
              CIP

"Fifty Years Out" adapted from a talk at the Harvard Medical School Reunion, Class of 1937, June 11–12, 1987; "Becoming a Doctor" adapted from a review of Melvin Konner's *Becoming a Doctor* in *The New York Review of Books,* August 13, 1987; "The Art and Craft of Memoir" adapted from "The Art and Craft of Memoir" series, The New York Public Library, February 11, 1986; "The Life in the Mind" adapted from a lecture at the Association of American Physicians Centennial Dinner, May 5, 1986; "In Time of Plague" adapted from "In Time of Plague: The History and Social Consequences of Lethal Epidemic Disease," a conference of The New School for Social Research, January 15–16, 1988; "AIDS and Drug Abuse" adapted from "AIDS and Drug Abuse: The Future," presented at the Business Council Meeting, The Homestead, Hot Springs, Virginia, October 10, 1987; "The Odds on Normal Aging" adapted from Brookhaven Symposia in Biology No. 33, September 30–October 3, 1984; "Obligations" adapted from Elihu Root Lecture II, "Human Health and Foreign Policy," published in *Foreign Affairs* 62:4, 976–987, 1984; "Comprehending My Cat Jeoffry" adapted from "The Awareness of Nature" the first Marie Nyswander Memorial Lecture, Beth Israel Medical Center, April 23, 1987; "Science and the Health of the Earth" adapted from Elihu Root Lecture III, published in *Foreign Affairs* 62:4, 987–994, 1984; "Cooperation" adapted from Lipkin Lectures, Museum of Natural History, February 1984; "Communication" adapted from Lipkin Lectures, Museum of Natural History, February 1984, published in *Missouri Review,* Spring 1992; "Connections" adapted from "The New Transnational Structure of Basic Science: Prospects and Apprehensions," The Medawar Lecture, 1990.

To Beryl and Abby, interested parties

Contents

Foreword

*I*n the early 1970s a remarkable thing happened to the *New England Journal of Medicine*. A series of occasional articles appeared with the modest title "Notes of a Biology Watcher." Written for physicians, they were a reminder that human beings are part of a natural world of subtlety and beauty. When the "Notes" were made available to the general public in two books, *The Lives of a Cell* and *The Medusa and the Snail*, their presence changed the perceptions and enhanced the lives of millions of readers.

Lew Thomas and I have certain superficial things in common: we're native New Yorkers, we attended the same medical school, and we interned at that crucible of life and death, the Boston City Hospital. From there Lew went on to become an outstanding biomedical investigator; professor of pediatrics, pathology, medicine, and biology; dean of two medical schools; and president of a great cancer center. What is this formidable figure really like? When I returned to New York after thirty peripatetic years, the most striking thing about Lew was his conversation, which was just as elegant as his writing.

On my recent move to the world of publishing, the first author

I contacted was Lewis Thomas. I suggested that he write a book
on how biomedical science has changed clinical practice since his
internship in the late 1930s when, as described in *The Youngest
Science*, "medicine made little or no difference." He found the
subject so compelling that he signed on to examine the advances
in biomedical science in the latter half of this century. At this
time of his life, however, the task turned out to be too massive
for any one man. With characteristic generosity he asked me if
I would have any interest in seeing approximately sixty unpub-
lished essays and talks he had prepared over the last ten years.
Four large portfolios soon arrived. As I went through them I
was again seized by the wonder of his writing. I then became
aware of a series of themes running through these essays: a
man summing up his love for his profession; his interest in some
of the major medical issues of our time—AIDS, drug abuse, and
aging; his concern for the planet Earth, "hanging there in space
and obviously alive"; and his solution, rooted in biology, to the
possible destruction of "so lovely a creature."

We owe a particular debt to Robert Stewart, premier editor
at Scribners. Among other things, it was he who suggested to
Lew the title of this book, *The Fragile Species*, which came from
one of its essays: "I am a member of a fragile species, still new
to the earth, the youngest creatures of any scale, here only a few
moments as evolutionary time is measured, a juvenile species, a
child of a species. We are only tentatively set in place, error-
prone, at risk of fumbling, in real danger at the moment of leaving behind only a thin layer of our fossils. . . ."

Most of Lew's books have no introduction, nor do they need
one. When he asked me to write a foreword for *The Fragile
Species* I was pleased but dismayed. How can one write gracefully
about a man of grace who has the capacity to continually amaze?
My best advice is to read these essays, in which the wonderful
things Lewis Thomas says are so strangely and beautifully inter-
twined with the way in which he says them.

KENNETH S. WARREN

Acknowledgments

*T*he writer is indebted to the dean and faculty of Cornell University Medical College for their generous hospitality during the period when, as scholar in residence, he gathered the material for this book. Indebted also, for the most obvious reasons, to Kenneth Warren and Robert Stewart.

Acknowledgment is also due to my accomplished secretary, Stephanie Hemmert, for having kept flawless track of so many pieces of paper needed for what follows.

Acknowledgments

I

Fifty Years
Out

Almost precisely fifty years ago, all of us passed a milestone in our intellectual development. Within my own memory it was more than a milestone; it was a monument of achievement. You will all remember it, even though it lasted for only a brief period, the span of time between the final examinations and the first weeks of interneship. It was that best of all possible times in our lives, the moment when we knew absolutely everything about everything. And, for most of us, certainly for me, it was the last moment of its kind in a professional lifetime.

Ever since, it has been one confusing ignorance after another, fifty years of knowing less and less about more and more, a full half century of advancing bewilderment about medicine, about disease mechanisms, about human society, about the economics of medicine, and a mixture of confusion and irritation with all the jargon terms invented to obscure our professional deficiencies—like, for instance, always referring to the Health Care System with its Providers and Consumers and Wellness (always in capital letters) replacing the old-fashioned doctors and patients and medicine and illness. And, to make it more of a fake, setting up acronyms like HMO, just as though Health Maintenance (in capital letters) were the easiest of things for doctors to do, let alone even a possibility. And cost-benefit analysis, and technology assessment. And holistic versus reductionist science in medicine. And ethics and morals, and the need in medical education for more humanities, let alone more humanity. As I recall, my father instructed me in the truth fifty years ago that any mention of medical ethics meant money, and morals meant sex, and that was that.

I cannot count the number of new items of ignorance I've picked up in fifty years; the list is simply too long. Instead, I have prepared another kind of list, shorter, more personally humbling, of some things I think I might have been learning more about if I hadn't been so puzzled all those years by medicine itself. These are matters that I assume most other people my age comprehend nicely, and I never got round to studying.

The Federal Reserve System is at the top of my list. I've never known what it was or what it did or how it did it, and what is more I don't want to be told. The same goes for the stock market, and for the bond market, and the word processor (one of which I actually possess and am baffled by), and the internal combustion engine, and the universe, black holes, galactic mirrors, those other universes, and space-time. Most of all, space-time. I cannot get ahold of it.

I even have troubles of my own with evolutionary biology. Not first principles, mind you, not the big picture, mostly just the details. I understand about randomness and chance, and selection, and adaptation, and all that, and I now know better than to talk, ever, about progress in evolution, never mind purpose. My problems come when I think about the earliest form of known life, those indisputable bacterial cells in rocks 3.7 billion years old, our Ur-grandparents for sure, then nothing but bacteria for the next two and one-half billion years, and now the chestnut tree in my backyard, my Abyssinian cat Jeoffry, the almost-but-not-quite free-living microbes living in all our cells disguised as mitochondria, and, just by the way, our marvelous, still-immature, dangerous selves, brainy enough to menace all nature unless distracted by music. We need a better word than chance, even pure chance, for that succession of events, while still evading any notion of progress. But to go all the way from a clone of archaebacteria, in just 3.7 billion years, to the B-Minor Mass and the Late Quartets, deserves a better technical term for the record than randomness.

I like the word *stochastic* better, because of its lineage in our language. The first root was *stegh*, meaning a pointed stake in the Indo-European of 30,000 years ago. *Stegh* moved into Greek as *stokhos*, meaning a target for archers, and then later on, in our language, targets being what they are and aiming arrows being as fallible as it is, *stokhos* was adapted to signify aiming and missing, pure chance, randomness, and thus stochastic. On that philological basis, then, I'm glad to accept all of evolution in a swoop, but I'm still puzzled by it.

The last thing on my short list is the phenomenon of aging, in which I'm beginning to take a close, diligent interest. Someone asked Bruce Bliven, the editor of *The New Republic*, then in his seventies, what it felt like to be an old man. Bliven retorted, "I don't feel like an old man. I feel like a young man with something the matter with him."

I've read some nice technical explanations for the process of aging: entropy and that business of thermodynamics and how all clocks run down, molecular errors piling up, all that. My current favorite is in an issue of the *Quarterly Review of Biology*, in an article titled "Thermal Noise and Biological Information," by H. A. Johnson. The notion is that the intrinsic noise produced by the sheer heat needed for living causes a steady degradation in the quality of the information on which living cells and their working parts depend for a living. According to the notion, the optimal temperature of 37° C, which enables enzymes to function at their best, is far too hot for the long-term preservation of information systems, and in the long run, say seventy-five years—which is not to say a lifetime—the corruptive effect of heat gradually overwhelms its activating function. We would all have been better off at room temperature, or poikilothermic like turtles, or best of all absolutely immortal by living at absolute zero Kelvin.

But then, of course, if it had been arranged that way, we'd all still be alive forever but, in the nature of things, we would still be those same archaebacteria born 3.7 billion years ago, unable to make molecular errors, deprived of taking chances, and therefore never blundering into brains. Fifty years out of Harvard Medical School, that may be the nicest thing I've learned about the world.

Becoming
a Doctor

*D*octors, dressed up in one professional costume or another, have been in busy practice since the earliest records of every culture on earth. It is hard to think of a more dependable or enduring occupation, harder still to imagine any future events leading to its extinction. Other trades—goldsmithing, embalming, cathedral architecture, hexing, even philosophy—have had their ups and downs and times of vanishing, but doctoring has been with us since we stumbled into language and society, and will likely last forever, or for as long as we become ill and die, which is to say forever.

What is it that we expected from our shamans, millennia ago, and still require from the contemporary masters of the profession? To *do* something, that's what.

The earliest sensation at the onset of illness, often preceding the recognition of identifiable symptoms, is apprehension. Something has gone wrong, and a glimpse of mortality shifts somewhere deep in the mind. It is the most ancient of our fears. Something must be done, and quickly. Come, please, and help, or go, please, and find help. Hence, the profession of medicine.

You might expect that such a calling, with origins in deepest antiquity, would by this time have at hand an immense store of traditional dogma, volumes and volumes of it, filled with piece after piece of old wisdom, tested through the ages. It is not so. Volumes do exist, of course, but all of them are shiny new, and nearly all the usable knowledge came in a few months ago. Medical information does not, it seems, build on itself; it simply replaces structures already set in place, like the New York skyline. Medical knowledge and technical savvy are biodegradable. The sort of medicine that was practiced in Boston or New York or Atlanta fifty years ago would be as strange to a medical student or interne today as the ceremonial dance of a !Kung San tribe would seem to a rock festival audience in Hackensack.

I take it further. The dilemma of modern medicine, and the underlying central flaw in medical education and, most of all, in the training of internes, is this irresistible drive to do something, anything. It is expected by patients and too often agreed to by their doctors, in the face of ignorance. And, truth to tell, ignorance abounds side by side with the neat blocks of precise scientific knowledge brought into medicine in recent years.

It is no new thing. In 1876, on the occasion of the country's first centennial, a book entitled *A Century of American Medicine, 1776–1876*, was published. The five authors were indisputable authorities in their several fields, from the faculties of Harvard, Columbia, and Jefferson Medical College. The book is a summary of the major achievements in American medicine of the previous century. The optimistic last sentence in the book is perhaps more telling than the writers may have realized: "It is better to have a future than a past." A very large part of the past in that century of medicine was grim.

Early on, there was no such thing as therapeutic science, and beyond the efforts by a few physicians to classify human diseases and record the natural history of clinical phenomena, no sort of reliable empirical experience beyond anecdotes. Therapeutics was a matter of trial and error, with the trials based on guesswork and the guesses based mostly on a curious dogma inherited down the preceding centuries from Galen. Galen himself (c. 130– c. 200) had guessed wildly, and wrongly, in no less than five hundred treatises on medicine and philosophy, that everything about human disease could be explained by the misdistribution of "humors" in the body. Congestion of the various organs was the trouble to be treated, according to Galen, and by the eighteenth century the notion had been elevated to a routine cure-all, or anyway treat-all: remove the excess fluid, one way or another. The ways were direct and forthright: open a vein and take away a pint or more of blood at a sitting, enough to produce faintness and a bluish pallor, place suction cups on the skin to draw out lymph, administer huge doses of mercury or various plant extracts to cause purging, and if all else failed induce vomiting. George Washington perhaps died of this therapy at the age of sixty-six. Hale and hearty, he had gone for a horseback ride in the snow, later in the day had a fever and a severe sore throat, took to his bed, and called in his doctors. His throat was wrapped in poultices, he was given warm vinegar and honey to gargle, and over the next two days he was bled from a vein for about five pints of blood. His last words to his physician were, "Pray take no more trouble about me. Let me go quietly."

Beginning around the 1830s, medicine looked at itself critically, and began to change. Groups of doctors in Boston, Paris, and Edinburgh raised new questions, regarded as heretical by most of their colleagues, concerning the real efficacy of the standard treatments of the day. Gradually, the first example of science applied to clinical practice came somewhat informally into existence. Patients with typhoid fever and delirium tremens, two of the most uniformly fatal illnesses of the time, were divided into two groups. One was treated by bleeding, cupping, purging, and the other athletic feats of therapy, while the other group received nothing more than bed rest, nutrition, and observation. The results were unequivocal and appalling, and by the mid-nineteenth century medical treatment began to fall out of fashion and the era known as "therapeutic nihilism" was well launched.

The great illumination from this, the first revolution in medical practice in centuries, was the news that there were many diseases that are essentially self-limited. They would run their predictable course, if left to run that course without meddling, and, once run, they would come to an end and certain patients would recover by themselves. Typhoid fever, for example, although an extremely dangerous and potentially fatal illness, would last for five or six weeks of fever and debilitation, but at the end about 70 percent of the patients would get well again. Lobar pneumonia would run for ten to fourteen days and then, in lucky, previously healthy patients, the famous "crisis" would take place and the patients would recover overnight. Patients with the frightening manifestations of delirium tremens only needed to be confined to a dark room for a few days, and then were ready to come out into the world and drink again. Some were doomed at the outset, of course, but not all. The new lesson was that treating them made the outcome worse rather than better.

It is difficult to imagine, from this distance, how overwhelming this news was to most physicians. The traditional certainty had been that every disease was aimed toward a fatal termination, and without a doctor and his energetic ministrations, or barring miraculous intervention by a higher force, all sick people would die of their disease. To recognize that this was not so, and that with rare exceptions (rabies the most notable one) many sick people could get well by themselves, went against the accepted belief of the time. It took courage and determination, and time, to shake off the old idea.

Looking back over the whole embarrassing record, the historians of that period must be hard put to it for explanations of the steadily increasing demand, decade after decade, for more doctors, more clinics and hospitals, more health care. You might think that people would have turned away from the medical profession, or abandoned it. Especially since, throughout the last half of the nineteenth century and the full first third of this one, there was so conspicuously little that medicine had to offer in the way of effective drugs or indeed any kind of technology. Opium, digitalis, quinine, and bromides (for the "nerves") were the mainstays. What else did physicians do during all those years that kept their patients calling and coming?

Well, they did a lot of nontechnology, and it was immensely effective. Mainly, they made diagnoses, explained matters to the patient and family, and then stood by, taking responsibility. To be sure, there were skeptics and critics all around, but they had always been around. Montaigne wrote bluntly, concerning doctors: "I have known many a good man among them, most worthy of affection. I do not attack them, but their art. It is only fear of pain and death, and a reckless search for cures, which blinds us. It is pure cowardice that makes us so gullible." Molière made delightful fun of doctors in his century. Dickens had some affection but no great respect for the doctors, most of them odd, bumbling eccentrics, who turned up as minor but essential characters in all his novels. Shaw was a scathing critic of medicine and its pretensions, clear into modern times.

But the public regard, and loyalty, somehow held. It is exemplified by a memorial tablet in the north wall of St. James Church in Piccadilly, in honor of Sir Richard Bright (1789–1858), the discoverer of the kidney disease which still bears his name, and a not-atypical Harley Street practitioner during the period of transition from the try-anything to the just-observe schools of medicine. The plaque reads, in part,

> Sacred to the memory of Sir Richard Bright, M.D. D.C.L.
> Physician Extraordinary to the Queen
>
> He Contributed to Medical Science Many Scientific Discoveries
> And Works of Great Value
> And Died While In the Full Practice of His Profession
> After a Life of Warm Affection
> Unsullied Purity
> And Great Usefulness

This is what nineteenth-century people expected their doctors to be, and believed most of them were in real life. The expectation survives to this day, but the reality seems to have undergone a change, in the public mind anyway.

There are many very good physicians around, as gifted and sought after as Bright was in his time, unquestionably better equipped by far to deal with life-threatening illnesses, trained to a level of comprehension of disease mechanisms beyond any

nineteenth-century imagination, but "warm affection" and "unsul-
lied purity" have an anachronistic sound these days, and even
"great usefulness" is open to public questioning. The modern
doctor is literally surrounded by items of high technology capable
of preventing or reversing most of the ailments that used to kill
people in their youth and middle years—most spectacularly, the
bacterial and viral infections chiefly responsible for the average
life expectancy of less than forty-five years in Bright's day. But
medicine's agenda still contains a long list of fatal or incapacitating
diseases, mostly the chronic disabilities of older people, and
there is still no technology for these, not even yet a clear understanding of their underlying mechanisms.

The unequivocal successes include miliary tuberculosis, tertiary syphilis of the brain and heart, poliomyelitis, the childhood
contagions, septicemias, typhoid, rheumatic fever and valvular
heart disease, and most of the other great infectious diseases,
now largely under control or already conquered. This was the
result of the second big transformation in medicine, starting
about fifty years ago with the introduction of the sulfonamides,
penicillin, and the other antibiotics, gifts straight from science.
The revolution continues in full force, thanks to what is now
called the "biological revolution," but it is still in its early stages.
With new technologies of fantastic power, such as recombinant
DNA and monoclonal antibodies, disease mechanisms that were
blank mysteries, totally inaccessible just a few years back, are
now at least open to direct scrutiny in detail. The prospects for
comprehending the ways in which cancer works, as well as other
illnesses on what is becoming a long list, are now matters of
high confidence and excitement among the younger researchers
within the universities and in industrial laboratories.

But the future is not yet in sight, and medicine is still stuck,
for an unknowable period, with formidable problems beyond the
reach of therapy or prevention. The technologies for making an
accurate diagnosis have been spectacularly effective, and at the
same time phenomenally complex and expensive. This new activity is beginning to consume so much of the time of the students
and internes, and the resources of the hospitals in which they

do their work, that there is less and less time for the patient. Instead of the long, leisurely ceremony of history-taking, and the equally long ritual of the complete physical examination, and then the long explanations of what has gone wrong and a candid forecast of what may lie ahead, the sick person perceives the hospital as an enormous whirring machine, with all the professionals— doctors, nurses, medical students, aides, and porters—out in the corridors at a dead run. Questionnaires, fed into computers along with items analyzing the patient's financial capacity to pay the bills, have replaced part of the history. Blood samples off to the laboratory, the CAT scan, and Nuclear Magnetic Resonance machines are relied upon as more dependable than the physical examination.

Everyone, even the visitors, seems pressed for time. There is never enough time, the whole place is overworked to near collapse, out of breath, bracing for the next irremediable catastrophe—the knife wounds in the emergency ward, the flat lines on the electroencephalogram, the cardiac arrests, and always everywhere on every ward and in every room the dying. The Hippocratic adage "Art is long, Life is short" is speeded up to a blur.

Everyone is too busy, urgently doing something else, and there is no longer enough time for the old meditative, speculative ward rounds or the amiable conversations at bedside. The house staff, all of them—internes, residents, junior fellows in for the year on NIH training fellowships—are careening through the corridors on their way to the latest "code" (the euphemism for the nearly dead or the newly dead, who too often turn out to be, in the end, the same), or deciphering computer messages from the diagnostic laboratories, or drawing blood and injecting fluids, or admitting in a rush the newest patient. The professors are elsewhere, trying to allocate their time between writing out their research requests (someone has estimated that 30 percent of a medical school faculty's waking hours must be spent composing grant applications), doing or at least supervising the research in their laboratories, seeing their own patients (the sustenance of a contemporary clinical department has become significantly de-

pendent on the income brought in by the faculty's collective private practice), and worrying endlessly about tenure (and
parking). About the only professionals who are always on the
wards, watching out for the unforeseen, talking and listening to
the patients' families, are the nurses, who somehow manage,
magically, to hold the place together for all its tendency to drift
toward shambles.

I have only two proposals, more like obsessive wishes for the
future than a recipe for the present. My first hope is for removal
of substantial parts of the curriculum in the first two years, making enough room for a few courses in medical ignorance, so that
students can start out with a clear view of the things medicine
does not know. My second hope is for more research into the
mechanisms of that still-unsolved list of human diseases. The
trouble with medicine today is that we simply do not know
enough, we are still a largely ignorant profession, faced by an
array of illnesses which we do not really understand, unable to
do much beyond trying to make the right diagnosis, shoring
things up whenever we can by one halfway technology or another
(the transplantation of hearts, kidneys, livers, and lungs are the
only measures available when we lack any comprehension of the
events responsible for the prior destruction of such organs). A
great deal of the time and energy expended in a modern hospital
is taken up by efforts to put off endgame.

We will be obliged to go on this way and at steadily increasing
expense, as far as I can see, until we are rid of disease—at least
rid of the ailments which now dominate the roster and fill the
clinics and hospitals. This is not asking for as much as it sounds.
We will never be free of our minor, self-limited ills, nor should
we be planning on postponing dying beyond the normal human
span of living—the seventies and eighties for most of us, the
nineties for the few more (or less) lucky among us. But there
is a great deal we will be able to do as soon as we have learned
what to do, both for curing and preventing. It can never be done
by guessing, as the profession learned in earlier centuries. Nor
can very much be changed by the trendy fashions in changed
"life-styles," all the magazine articles to the contrary; dieting,

jogging, and thinking different thoughts may make us feel better while we are in good health, but they will not change the incidence or outcome of most of our real calamities. We are obliged, like it or not, to rely on science for any hope of solving such biological puzzles as Alzheimer's disease, schizophrenia, cancer, coronary thrombosis, stroke, multiple sclerosis, diabetes, rheumatoid arthritis, cirrhosis, chronic nephritis, and now, topping the list, AIDS. When we have gained a clear comprehension, in detail, of what has gone wrong in each of these, medicine will be earning its keep, in spades.

The Art and Craft
of Memoir

*I*t should be easier, certainly shorter work, to compose a memoir than an autobiography, and surely it is easier to sit and listen to the one than to the other. An autobiography, I take it, is a linear account of one thing after another leading, progressively one hopes, to one's personal state of affairs at the moment of writing. In my case, this would run to over seventy years, one after the other, discounting maybe twenty-five of the seventy spent sleeping—leaving around forty-five to be dealt with. Even so, a lot of time to be covered if all the events were to be recalled and laid out. But discount again the portion of those 16,500 days, 264,000 waking hours spent doing not much of anything, reading the papers, staring at blank sheets of paper, walking from one room to the next, speaking a great deal of small talk and listening to still more, waiting around for the next thing to happen. Delete all this as irrelevant, then line up what's left in the proper linear order without fudging. There you are with an autobiography, now relieved of an easy three-fourths of the time lived, leaving only eleven years, or 4,000 days, or 64,000 hours. Not much to remember, but still too much to write down.

But now take out all the blurred memories, all the recollections you suspect may have been dressed up by your mind in your favor, leaving only the events you can't get out of your head, the notions that keep leaping to the top of your mind, the ideas you're stuck with, the images that won't come unstuck including the ones you'd just as soon do without. Edit these down sharply enough to reduce 64,000 hours to around thirty minutes, and there's your memoir.

In my case, going down this shortened list of items, I find that most of what I've got left are not real memories of my own experience, but mainly the remembrances of other people's thoughts, things I've read or been told, metamemories. A surprising number turn out to be wishes rather than recollections, hopes that the place really did work the way everyone said it was supposed to work, hankerings that the one thing leading to another has a direction of some kind, and a hope for a pattern from the jumble, an epiphany out of entropy.

To begin personally on a confessional note, I was at one time, at my outset, a single cell. I have no memory of this stage of my life but I know it to be true because everyone says so. There was of course a sort of half-life before that, literally half, when the two half-endowed, haploid gametes, each carrying half my chromosomes, were off on their own looking to bump into each other and did so, by random chance, sheer luck, for better or worse, richer or poorer, et cetera, and I got under way.

I do not remember this, but I know that I began dividing. I have probably never worked so hard, and never again with such skill and certainty. At a certain stage, very young, a matter of hours of youth, I sorted myself out and became a system of cells, each labeled for what it was to become—brain cells, limbs, liver, the lot—all of them signaling to each other, calculating their territories, laying me out. At one stage I possessed an excellent kidney, good enough for any higher fish, then I thought better and destroyed it all at once, installing in its place a neater pair for living on land. I didn't plan on this when it was going on, but my cells, with a better memory, did.

Thinking back, I count myself lucky that I was not in charge

at the time. If it had been left to me to do the mapping of my
cells I would have got it wrong, dropped something, forgotten
where to assemble my neural crest, confused it. Or I might have
been stopped in my tracks, panicked by the massive deaths,
billions of my embryonic cells being killed off systematically in
order to make room for their more senior successors, death on
a scale so vast I cannot think of it without wincing. By the time
I was born, more of me had died than survived. It is no wonder
I cannot remember; during that time I went through brain after
brain for nine months, finally contriving the one model that could
be human, equipped for language.

It is because of language that I am able now to think farther
back into my lineage. By myself, I can only remember two parents, one grandmother, and the family stories of Welshmen back
into the shadows when all the Welsh were kings, but no further.
From there on I must rely on reading the texts.

They instruct me that I go back to the first of my immediate
line, the beginner, the earliest *Homo sapiens*, human all the way
through, or not quite human if you measure humanness as I do
by the property of language and *its* property, the consciousness
of an indisputable singular, unique self. I am not sure how far
back that takes me, and no one has yet told me about this
convincingly. When did my relations begin speaking?

Writing is easier to trace, having started not more than a few
years back, maybe 10,000 years, not much more. Tracking
speech requires guesswork. If we were slow learners, as slow
as we seem to be in solving today's hard problems, my guess is
that we didn't begin talking until sometime within the last
100,000 years, give or take 50,000. That is what's called a rough
scientific guess, but no matter, it is an exceedingly short time
ago and I am embarrassed at the thought that so many of my
ancestors, generations of them—all the way back to the very
first ones a million-odd years ago—may have been speechless.
I am modestly proud to have come from a family of tool-makers,
bone-scratchers, grave-diggers, cave-painters. Humans all. But
it hurts to think of them as so literally dumb, living out their
lives without metaphors, deprived of conversation, even small

talk. I would prefer to have had them arrive fully endowed, talking their heads off, the moment evolution provided them with brain cases large enough to contain words, so to speak. But it was not so, I must guess, and language came late. I will come back to this matter.

What sticks in the top of my mind is another, unavoidable aspect of genealogy, far beyond my memory, but remembered still, I suspect, by all my cells. It is a difficult and delicate fact to mention. To face it squarely, I come from a line that can be traced straight back, with some accuracy, into a near infinity of years before my first humanoid ancestors turned up. I go back, and so do you, like it or not, to a single Ur-ancestor whose remains are on display in rocks dated approximately 3.7 thousand million years ago, born a billion or so years after the earth itself took shape and began cooling down. That first of the line, our n-granduncle, was unmistakably a bacterial cell.

I cannot get this out of my head. It has become, for the moment, the most important thing I know, the obligatory beginning of any memoir, the long-buried source of language. We derive from a lineage of bacteria, and a very long line at that. Never mind our embarrassed indignation when we were first told, last century, that we came from a family of apes and had chimps as near cousins. That was relatively easy to accommodate, having at least the distant look of a set of relatives. But this new connection, already fixed by recent science beyond any hope of disowning the parentage, is something else again. At first encounter, the news must come as a kind of humiliation. Humble origins, indeed.

But then it is some comfort to acknowledge that we've had an etymological hunch about such an origin since the start of our language. Our word "human" comes from the proto-Indo-European root *dhghem*, meaning simply "earth." The most telling cognate word is "humus," the primary product of microbial industry. Also, for what it is worth, "humble." Also "humane." It gives a new sort of English, in the sense of a strange spin, to the old cliché for an apology: "Sorry, I'm only human."

Where did that first microorganism, parent of us all, come

from? Nobody knows, and in the circumstance it is anyone's guess, and the guesses abound. Francis Crick suggests that the improbability of its forming itself here on earth is so high that we must suppose it drifted in from outer space, shifting the problem to scientists in some other part of the galaxy or beyond. Others assert that it happened here indeed, piecing itself together molecule by molecule, over a billion years of chance events under the influence of sunlight and lightning, finally achieving by pure luck the exactly right sequence of nucleotides, inside the exactly right sort of membrane, and we were on our way.

No doubt the first success occurred in water. And not much doubt that the first event, however it happened, was the only such event, the only success. It was the biological equivalent of the Big Bang of the cosmophysicists, very likely a singular phenomenon, a piece of unprecedented good luck never to be repeated. If the sheer improbability of the thing taking place more than once, spontaneously and by chance, were not enough, consider the plain fact that all the cells that came later, right up to our modern brain cells, carry the same strings of DNA and work by essentially the same genetic code. It is the plainest evidence of direct inheritance from a single parent. We are all in the same family—grasses, seagulls, fish, fleas, and voting citizens of the republic.

I ought to be able to remember the family tie, since all my cells are alive with reminders. In almost everything they do to carry me along from one day to the next, they use the biochemical devices of their microbial forebears. Jesse Roth and his colleagues at NIH have shown that the kingdom of bacteria had already learned, long before nucleated cells like ours came on the scene, how to signal to each other by chemical messages, inventing for this purpose molecules like insulin and a brilliant array of the same peptides that I make use of today for instructing my brain cells in proper behavior.

More than this, I could not be here, blinking in the light, without the help of an immense population of specialized bacteria that swam into cells like mine around a billion years ago and stayed there, as indispensable lodgers, ever since, replicating on their own, generation after generation. These are my mitochondria, the direct descendants of the first bacteria that learned how to make use of oxygen for energy. They occupy all my cells, swarming from one part to another wherever there is work to do. I could not lift a finger without them, nor think a thought, nor can they live without me. We are symbionts, my mitochondria and I, bound together for the advance of the biosphere, living together in harmony, maybe even affection. For sure, I am fond of my microbial engines, and I assume they are pleased by the work they do for me.

Or is it necessarily that way, or the other way round? It could be, I suppose, that all of me is a sort of ornamented carapace for colonies of bacteria that decided, long ago, to make a try at real evolutionary novelty. Either way, the accommodation will do.

The plants are in the same situation. They have the same swarms of mitochondria in all their cells, and other foreign populations as well. Their chloroplasts, which do the work of tapping solar energy to make all sugar, are the offspring of ancient pigmented microorganisms called cyanobacteria, once known as blue-green algae. These were the first creatures to learn—at least 2.5 billion years ago—how to use carbon dioxide from the air, and plain water, and sunlight, to manufacture food for the market.

I am obsessed by bacteria, not just my own and those of the horse chestnut tree in my backyard, but bacteria in general. We would not have nitrogen for the proteins of the biosphere without the nitrogen-fixing bacteria, most of them living like special tissues in the roots of legumes. We would never have decay; dead trees would simply lie there forever, and so would we, and nothing on earth would be recycled. We could not keep cows, for cattle cannot absorb their kind of food until their intestinal bacteria have worked it over, and for the same reason there would be no termites to cycle the wood; they are, literally, alive with bacteria. We would not have luminous fish for our aquariums, for the source of that spectacular light around their eyes is their private colonies of luminescent bacteria. And we would never have obtained oxygen to breathe, for all the oxygen in our air is exhaled for our use by the photosynthetic microbes in the upper waters of the seas and lakes, and in the leaves of forests.

It was not that we invented a sophisticated new kind of cell with a modern nucleus and then invited in the more primitive and simpler forms of life as migrant workers. More likely, the whole assemblage came together by the joining up of different kinds of bacteria; the larger cell, the original "host," may have been one that had lost its rigid wall and swelled because of this defect. Lynn Margulis has proposed that the spirochetes were part of the original committee, becoming the progenitors of the cilia on modern cells, also the organizers of meiosis and mitosis, the lining up of chromosomes, the allocation of DNA to progeny—in effect, the reading of all wills. If she is right about this, the spirochetes were the inventors of biological sex and all that, including conclusive death.

The modern cell is not the single entity we thought it was a few years ago. It is an organism in its own right, a condominium, run by trustees.

If all this is true, as I believe it to be, the life of the earth is more intimately connected than I used to think. This is another thing on my mind, so much in my head these days that it crowds out other thoughts I used to have, making me sit up straight now, bringing me to my feet and then knocking me off them. The world works. The whole earth is alive, all of a piece, one living thing, a creature.

It breathes for us and for itself; and what is more, it regulates the breathing with exquisite precision. The oxygen in the air is not placed there at random, any old way; it is maintained at precisely the optimal concentration for the place to be livable. A few percentage points more than the present level and the forests would burst into flames, a few less and most life would strangle. It is held there, constant, by feedback loops of information from the conjoined life of the planet. Carbon dioxide, inhaled by the plants, is held at precisely the low level which would be wildly improbable on any lifeless planet. And this happens to be the right concentration for keeping the earth's temperature, including the heat of the oceans, exactly right. Methane, almost all of it the product of bacterial metabolism, contributes also to the greenhouse effect, and methane is held steady. Statesmen must keep a close eye on the numbers these days; we are already pushing up the level of CO_2 by burning too much fuel and cutting too much forest, and the earth may be in for a climate catastrophe within the next century.

But there it is: except for our meddling, the earth is the most stable organism we can know about, a complex system, a vast intelligence, turning in the warmth of the sun, running its internal affairs with the near infallibility of a huge computer. Not entirely infallible, however, on the paleontological record. Catastrophes occur by nature, crashes, breakdowns in the system: ice ages, meteor collisions, volcanic eruptions, global clouding, extinctions of great masses of its living tissue. It goes *down*, as we say of computers, but never out, always up again with something new to display to itself.

The newest of all things, the latest novelty among its working parts, seems to be us—language-speaking, song-singing, tool-making, fire-warming, comfortable, warfaring mankind, and I am of that ilk.

I cannot remember anything about learning language as a child. I do have a few memories of studying to read and write, at age four or five, I think, but I have no earlier recollection at all of learning speech. This surprises me. You'd think the first word, the first triumphant finished sentence, would have been such a stunning landmark to remain fixed in memory forever, the biggest moment in life. But I have forgotten. Or perhaps it never embedded itself in my mind. Being human, I may have known all along about language, from the time of my first glimpse of human faces, and speech just came, as natural a thing to do as breathing. The reason I cannot remember the learning process, the early mistakes, may be that at that time they were not mistakes at all, just the normal speech of childhood, no more memorable than the first drawn breath.

All my adult life I have hoped to speak French one day like a

Frenchman, but I am near to giving up, troubled. Why should any small French child, knee-high, be able to do so quickly something that I will never learn to do? Or, for that matter, any English or Turkish child living for a few months in Paris? I know the answer, but I don't much like to hear it, implying as it does that there are other knacks that I have lost as well. Childhood is the time for language, no doubt about it. Young children, the younger the better, are good at it; it is child's play. It is a onetime gift to the species, withdrawn sometime in adolescence, switched off, never to be regained. I must have had it once and spent it all on ordinary English. I possessed a splendid collection of neurons, nested in a center somewhere in my left hemisphere, probably similar to the center in a songbird's brain—also on his left side—used for learning the species' song while still a nestling. Like mine, the bird's center is only there for studying in childhood; if he hears the proper song at that stage he will have it in mind for life, ornamenting it later with brief arpeggios so that it becomes his own, particular, self-specific song, slightly but perceptibly different from the song of all his relatives. But if he cannot hear it as a young child, the center cannot compose it on its own, and what comes out later when he is ready for singing and mating is an unmelodious buzzing noise. This is one of the saddest tales in experimental biology.

Whatever happened in the human brain to make this talent for language a possibility remains a mystery. It might have been a mutation, a new set of instructions in our DNA for the construction of a new kind of center, absent in all earlier primates. Or it could have been a more general list of specifications: i.e., don't stop now, keep making more columnar modules of neurons, build a bigger brain. Perhaps any brain with a rich-enough cortex can become a speaking brain, with a self-conscious mind.

It is a satisfying notion for a memoir. I come from ancestors whose brains evolved so far beyond those of all their relatives that speech was the result, and with this in hand they became the masters of the earth, God's image, self-aware, able to remember generations back and to think generations ahead, able to write things like "in the beginning was the word." Nothing lies any longer beyond reach, not even the local solar system or out into the galaxy and even, given time, beyond that for colonizing the Universe. In charge of everything.

But this kind of talk is embarrassing; it is the way children talk before they've looked around. I must mend the ways of my mind. This is a very big place, and I do not know how it works, nor how I fit in. I am a member of a fragile species, still new to the earth, the youngest creatures of any scale, here only a few moments as evolutionary time is measured, a juvenile species, a child of a species. We are only tentatively set in place, error-prone, at risk of fumbling, in real danger at the moment of leaving behind only a thin layer of our fossils, radioactive at that.

With so much more to learn, looking around, we should be more embarrassed than we are. We are different, to be sure, but not so much because of our brains as because of our discomfiture, mostly with each other. All the other parts of the earth's life seem to get along, to fit in with each other, to accommodate, even to concede when the stakes are high. They live off each other, devour each other, scramble for ecological niches, but always within set limits, with something like restraint. It is a rough world, by some of our standards, but not the winner-take-all game that it seemed to us awhile back. If we look over our shoulders as far as we can see, all the way past trillions of other species to those fossil stromatolites built by enormous communities of collaborating microorganisms, we can see no evidence of meanness or vandalism in nature. It is, on balance, an equable, generally amiable place, good-natured as we say.

We are the anomalies for the moment, the self-conscious children at the edge of the crowd, unsure of our place, unwilling to join up, tending to grabbiness. We have much more to learn than language.

But we are not as bad a lot as some of us say. I do not agree with this century's fashion of running down the human species as a failed try, a doomed sport. At our worst, we may be going through the early stages of a species' adolescence, and everyone remembers what that is like. Growing up is hard times for an individual but sustained torment for a whole species, especially one as brainy and nervous as ours. If we can last it out, get through the phase, shake off the memory of this century, wait for a break, we may find ourselves off and running again.

This is an optimistic, Panglossian view, and I am quick to say that I could be all wrong. Perhaps we have indeed come our full evolutionary distance, stuck forever with our present behavior, as mature as we ever will be for as long as we last. I doubt it. We are not out of options.

I am just enough persuaded by the sociobiologists to believe that our attitudes toward each other are influenced by genes, and by more than just the genes for making grammar. If these alone were our only wired-in guides to behavior, we would be limited to metaphor and ambiguity for our most important messages to each other. I think we do some other things, by nature.

From earliest infancy on, we can smile and laugh without taking lessons, we recognize faces and facial expressions, and we hanker for friends and company. It goes too far to say that we have genes for liking each other, but we tend in that direction because of being a biologically social species. I am sure of that point: we are more compulsively social, more interdependent and more inextricably attached to each other than any of the celebrated social insects. We are not, I fear, even marginally so committed to altruism as a way of life as the bees or ants, but at least we are able to sense, instinctively, certain obligations to one another.

One human trait, urging us on by our nature, is the drive to be useful, perhaps the most fundamental of all our biological necessities. We make mistakes with it, get it wrong, confuse it with self-regard, even try to fake it, but it is there in our genes, needing only a better set of definitions for usefulness than we have yet agreed on.

So we are not entirely set in our ways. Some of us may have more dominant genes for getting along with others. I suspect, glancing around my life, that we are also endowed with other, inhibitory alleles, widely spread for the enhancement of anomie. Most of us are a mixture. If we like, we can sit tight, trusting nature for the best of possible worlds to come. Or we can hope for better breeding, in both senses of the term, as our evolution proceeds.

Our microbial ancestors made use of quicker ways for by-passing long stretches of evolutionary time, and I envy them. They have always had an abundance of viruses, darting from one cell to another across species lines, doing no damage most of the time ("temperate" viruses, as they are called), but always picking up odds and ends of DNA from their hosts and then passing these around, as though at a great party. The bits are then used by the recipients for their betterment, new tricks for coping with new contingencies.

I hope our species has a mechanism like this. Come to think of it, maybe we do. After all we live in a sea of our own viruses, most of which seem to be there for no purpose, not even to make us sick. We can hope that some of them might be taking hold of useful items of genetic news from time to time, then passing these along for the future of the race.

It makes a cheerful footnote, anyway: next time you feel a cold coming on, reflect on the possibility that you may be giving a small boost to evolution.

The Life in
the Mind

*T*he conventional way of looking at the human brain is to see it as an intricately wired calculating machine, receiving inputs of information from those regions of the outside world for which it has been structurally equipped with receptors, then either storing the information for later retrieval or sending it along from one center to another for immediate processing and action, depending on the circumstances. In recent years, the brain has become immensely more complicated. It is not just a hard-wired device, but turns out now to run its inner affairs by an exchange of chemical messages, hundreds of them, among its billions of neurons. It has become the custom in recent years to refer respectfully to this awesome device as the most complicated and elaborate of all structures in the universe, perhaps even including the universe itself; indeed, there are some who suggest that the universe itself would not exist without it; that the only reality is what the human brain perceives as reality, like that falling tree in that earless forest. Black holes only exist because we observe them into existence.

This view fits with the prevailing notion that all of nature is itself an immense machine, having started itself up some 3.7

thousand million years ago as an unaccountable single prokaryotic cell, maybe having been preceded by strands of RNA in possession of the enzyme-like properties recently assigned to certain configurations of RNA, finally learning how to make proper proteins and a proper cell membrane, and these things having been followed, much later on, by symbiotic arrangements among prokaryotes resulting in proper nucleated cells of our kind, and followed then, as the night the day, by the burst of metazoan life culminating, at long last, in all of us and all our proper brains.

The whole affair, then, is a machine. Nature is an immense mechanism, operating itself in accordance with the laws of physics. We, and our brains, are working parts of the machinery, having made our appearance here and having our existence because of the operation of those laws, set in place on what we like to see as the pinnacle by the beneficent operation of chance and quantum mechanics. Pure luck, indeterminate and intentionless, all the way.

This view takes us a long distance toward understanding our place in nature, but not quite the full distance. We are still stuck with the problem of consciousness, and because of this not-quite-settled matter, we are stuck as well with the incessant questions with which our consciousness continues to plague us and disturb our sleep (for which also, by the way, we do not have a good explanation). Questions like: Are we the only creatures on the whole planet with real consciousness? Why is being being; why not nonbeing? Why should there be something, instead of nothing? How do you organize a life, or a society, in accordance with physical laws that forbid purpose, causality, morality, and progress, especially when you have to do so with brains that stand alive with these very notions? Where's the fun in it?

The professional psychologists managed to evade discomfort, through the first half of this century and more, by taking up strict scientific attitudes early on and sticking to the facts at hand, to questions like: how does the human mind work? The prompt answer, based on the facts at hand, was there's no such thing. There is, to be sure, a brain, a machine, but everything it does is determined by whatever experiences it happens to have had, reflex-fashion, and anything beyond this—thought, for example—is in the realm of something disreputable called "mentalism," the professional term for fantasy, a waste of time for intellectual effort, especially when there is no such thing as intellectual effort, only more automatic reflexes to contingencies.

Pleasure, clear up through the 1950s, simply did not exist in the human mind. It was merely the sensation, or set of sensations, that resulted from the relief of one or another kind of pain. "Drives" were the thing. Drives produced tension, a painful condition, and as soon as a drive, whatever, was relieved by satiety, its absence became pleasure. This despite the ancient etymology of the words "satiety" and "satisfaction," which come straight down from the Indo-European root "sa," which meant sad. Freud believed that the only drive that mattered was the libido; because of this he thought that what young children seemed to have as pleasure was not the real article, something else recapitulating unsatisfiable infantile sexuality. The behaviorists had lists of other drives, but they were no better than Freud at accounting for fun, much less allowing it into their systems.

And then, in the early 1950s, a series of experiments on rats was begun by James Olds and his colleagues that should have changed everything about our view of the mind, but hasn't yet. I have been reading my way through the decades of neurobiology launched by Olds in 1953, and that is what I propose to talk about.

Like a good many other important scientific observations, this one was made by accident in the course of looking for something different. Olds was interested in using the techniques devised by W. R. Hess of Zurich for localizing various cerebral functions by stimulating electrodes implanted permanently in one part or another of the living animal brain. Magoun had shown earlier that the technique could be used for studying the sleep and wakefulness centers in the reticular substance of the lower midbrain, and James Olds intended to follow up on this work. By a stroke of luck, and an instrument bent slightly out of shape, his electrode ended up in a narrow bundle of fibers near the midline, long known to neuroanatomists as the medial forebrain bundle.

When this bundle was stimulated, the rats behaved in a very curious manner. They were free to run, but kept returning to the exact area in the experimental box where the first charge had been delivered, as though waiting for another. Olds, who must have been a master experimenter, promptly set up a battery of tests to see what this was all about. First, he ran the rats through a conventional maze and found that they quickly learned which channel to take in order to receive stimulation. Then he tested the desirability of the stimulus by arranging various barriers to interfere with the animals' access to the stimulus: he found that they would climb heights, circumvent blockades, and cross electrified grates that were painful enough to keep away ordinary rats in search of food. Finally, he adapted the Skinner box to permit his rats to stimulate their own brains by pressing a treadle.

Now, left to themselves, in charge of their own stimulation, the animals became engrossed. When the duration of each pulse and the magnitude of electrical current had been arranged to its satisfaction, a rat would remain at the treadle for hours and hours, day and night, pressing the lever as frequently as 10,000 times every hour. The world was given up. If food and water were withheld, and the hungry animal then presented with the choice of food or the treadle, he invariably chose the electrical charge in his medial forebrain bundle and left the food alone. Indeed, unless detached from the apparatus, the animals would stimulate themselves until totally exhausted and near to dying.

The phenomenon was unlike any other reward system previously studied. It seemed wholly unrelated to the drives to satisfy hunger or thirst, unconnected to sexual arousal or satisfaction. There seemed to be no such thing as satiety, nor did the response result in anything like genuine addiction. When the source of current was switched off the animals usually would try the lever a few times, then simply give it up and go to sleep.

Olds and his colleague Milner next set about learning whether the "pleasure center," as they termed the area being stimulated, was unique within the rat brain, or whether there were other places equally attractive. Their results launched a cascade of similar experiments by other neuroscientists around the world, and this aspect of the problem is still, more than three decades later, unsettled. The medial forebrain bundle is narrow and sharply localized, but it appears to have connections to almost every other part of the brain, from the prefrontal cortex to the pons. Most people now agree that similar, though less spectacular, responses can be obtained by stimulating electrodes placed in the frontal cortex, the hippocampus, the medial region of the thalamus, the anterior part of the locus ceruleus, the nucleus accumbens, and several other places. The fibers carrying the message are believed to belong to the noradrenergic rather than the dopaminergic system.

I don't know of any other single problem in neurobiology that has generated as much work, and as many papers in journals around the world, as this initial research by James Olds. The literature up through the late 1970s fills hundreds of lines in *Index Medicus*, and I have collected as much of this as my secretary and I have been able to locate in nearby libraries. It makes a formidable pile but seems to have leveled off in recent years. I have the impression, as an outsider, that the community involved in the problem has, at least for the time, nearly run out of questions to ask.

The rat is not alone in possessing a pleasure center or, better, centers. Similar experiments have been performed in rabbits, dogs, cats, goldfish, birds, several species of monkeys, and, to a necessarily limited extent, man. John Lilly reported sinking an electrode into the brain of a dolphin and eliciting the same behavior, which comes as a great surprise considering that the dolphin is, to begin with, or I have always thought so, as filled with pleasure as an animal can be. The outcome in each species has been essentially the same. There are tracts in the vicinity of the medial longitudinal bundle, and less responsive places in other locations, where stimulation causes a sensation that the animals like very much and wish to have repeated as frequently as possible. It is not the release of endorphin, since naloxone has no effect on the phenomenon. There are also other regions, some of them very close to those eliciting pleasure, that produce aversive reactions resembling the response to pain or danger, but these seem to be totally unrelated to the tracts conveying pleasure.

The range of animals possessing a pleasure center is already broad enough to allow the generalization that it is a property built into all vertebrate brains, although I haven't found a reference yet to reptiles. Nor do I know whether invertebrates have been excluded; I do hope someone is trying it out in lobsters and limulus. If I could just read somewhere that *Aplysia* has a string of neurons yielding pleasure, or that an earthworm or a drosophila can be induced to stimulate a particular set of ganglia, I would organize brass bands everywhere for dancing in all the streets of the world until I ran out of money.

I like to think that it is precisely because of the existence of loose-minded people like me that the neurobiologists have gone carefully with this problem. That, and perhaps also because of their hunch that this time they were onto something with very few if any imaginable applications for practical use. Interesting perhaps, a nice model for studying psychopharmacology—what drugs will enhance, which ones suppress the pleasure response? But something to go carefully with, maybe even something a bit scary: John Lilly pounding electrodes into the head of a dolphin, Robert Heath with his schizophrenic patients bristling with implants back in the 1950s and 1960s. Better play it quiet.

So the work went on with some intensity in the 1970s, mostly in rats and monkeys, but low-keyed. The term "pleasure center" was dropped by mutual consent all round; it no longer turns up in any of the neuroscience publications, nor in the *Index Medicus*. Once in a while it is mentioned in a historical review, but with deprecating remarks about the folly of anthropomorphizing. Whatever it is that goes on in the brain of a rat cannot be interpreted as having any meaning whatever for the human mind of man. Instead of pleasure, or pleasure center, or pleasure stimulation, the phenomenon is now referred to in all the articles as ICSS. If you want to find it, look in the index under Intracranial Self-Stimulation. ICSS, that's all it is. In the September 21,

1984, issue of *Science*, a remarkable issue devoted entirely to papers on the present status and future prospects of neurobiology, there is no mention of it.

Maybe it will vanish at last into the literature, reappearing from time to time or glimpsed staring from an attic window like a mad cousin of neurobiology, a sort of unaccountable embarrassment to the field. Or maybe it will suddenly catch fire again from a new idea and consume everyone's attention for another thirty years. Right now, though, it seems to be in the doldrums. A recent lively paper from the University of Pennsylvania describes an automated, computerized apparatus for analyzing a rat's behavior under ICSS around the clock, distinguishing quantitatively between the degrees of reward and other factors affecting the rat's behavior. Perhaps something like this will start things up again.

Meanwhile, it seems to me that the phenomenon has been laid in front of us as a free gift, taken as far as it can be taken by skilled, reductionist research, resting there for the mind's eye, free game for speculation by any interested party, including me.

Personally, I like it this way. I'm just as glad that the problem is stuck, for the time anyway, not likely to be swept along and explained away in the near future. I prefer that. For it has raised, it seems to me, one of the loveliest of all the possibilities raised by biological science in memory. Maybe the loveliest of all, considering the way most of us tend to look at nature and her works these days.

Namely, there is such a thing as pure pleasure, and there is a mechanism for mediating it alive in our brains. And not just our superior, world-class, super-primate human brains, but perhaps in all kinds of brains. Granted, this is a distorting, terribly unnatural, fundamentally misleading way for it to be revealed to us, by so artefactual a system for demonstrating its existence. There is something distasteful, even nasty, about viewing a rat nearly killing himself by stimulating a part of his brain that gives him ineffable pleasure. But pass that, and think of what lies behind these experiments, what has really been revealed. There is at least one tract, probably more, connected to many different parts of the brain, whose function it is to convey pleasure.

Pleasure in what? Is it simply built in, an autonomous mechanism deep in the brain, running on its own, always switched on, reaching consciousness only when other noises are shut off? Or does it depend for its sources on firing of receptors connected to the world outside the brain, like most other events in the central nervous system? Is it a sense of reward, only turned on by some *special* kind of news from outside? Not hunger or thirst, not pain or the relief of pain, not touch or proprioception, nothing to do with vision or sound, not sex, not domination, not territory, not language or song of any kind of sound, only something giving silent pleasure?

What could it be, then, this news that is constantly streaming into the central nervous system of all animals, traveling the right track, finding the right receptors, then sent along in relays of the same news to the limbic system and up into the frontal lobes, spreading, so to speak, the news around? It must be something important, signifying something we need to have signified for our satisfaction in life.

I have a vote in this, and I vote for living. I suggest that the medial forebrain bundle carries the impulses, coming in from cells all over the body, bearing the news that they are alive. Simple as that. I insert this notion here, not in order to gain priority and win a prize for myself, but because in the blessed lull of the research in its present phase I can stick the idea in before it is snatched away and destroyed by the forward progress of science. The sense of being alive *should* exist, even if it proves not to when the work has moved further along. Pleasure in being alive *ought* to exist as a special, independent, autonomous sense. If I had the responsibility for putting together a closed ecosystem as huge as the one on this planet, with the intention of having it persist and survive by evolution, I would put this one property in at the very beginning, along with ribosomes and cell membranes, as a basic property of everything alive, excluding it from natural selection and any sort of competition, violating all the rules but never mind. Never mind the rules in this single case, make an exception here, allow for the pure fun that ravens have swooping down in the winds along the sides of mountain cliffs, allow for what cats do when not busy with serious cat business, make a provision for humans, especially young children playing, put in a mechanism that can handle the inside of the messages conveyed both by the Fourteenth Quartet and the fourth movement of the *Missa Solemnis,* where the violin and the human voice suddenly turn into a single voice, and install the receptor for that word in that line of that poem, that jolt of that image. Take into account the need of an organism to know, for sure, that it is alive. In short, make the game worth playing, for all the players.

Come to think of it, I hope the work on ICSS doesn't move from where it now stands. It is exactly the right piece of scientific information, just where it now is, in this part of this century. Don't, for heaven's sake, move it; just leave it there for thinking about.

And don't talk to me about cocaine or amphetamine or the other pharmacological tricks that may have come into fashion for artificially switching on the pleasure mechanism. I doubt that they, or any other artificial devices, can really switch on something as fundamental as this. It is more likely always running, always switched on. If drugs are indeed involved, as indeed they may be, they are meddling with it, disturbing its true function out of reality, in a sense tormenting it as Olds's electrodes were really tormenting the medial forebrain bundle. James Thurber had it right: "Let your mind alone." Perhaps the best way to feel the pleasure, if you are willing to make the sacrifice, is to shut down most of what you usually keep your mind doing in order to get through the day.

Charles Darwin must have caught a glimpse of this, late in his day, when he wrote his saddest words: "Now for many years, I cannot endure to read a line of poetry. . . . I have also almost lost my taste for pictures or music. . . . My mind seems to have become a kind of machine for grinding general laws out of a large collection of facts. . . . The loss of those tastes is a loss of happiness." Technically, Darwin may have been suffering from an overfunctioning of the "gate-control system" for modulating incoming sensory information.

But Darwin's medial forebrain bundle had worked wonderfully well earlier in his life. In an 1858 letter to his wife, Emma, he wrote: "I fell asleep in the grass and awoke with a chorus of birds singing around me, and squirrels running up the trees . . . and I did not care one penny how the beasts or birds had been formed." It may be that if he had not had his own mechanism for inner pleasure tuned up and working to perfection early in his life, we might never have had *The Origin of Species*. Indeed, if it were not for the medial forebrain bundle, and the messages carried along its fibers, whatever they are, we might never have become the species we hope to be.

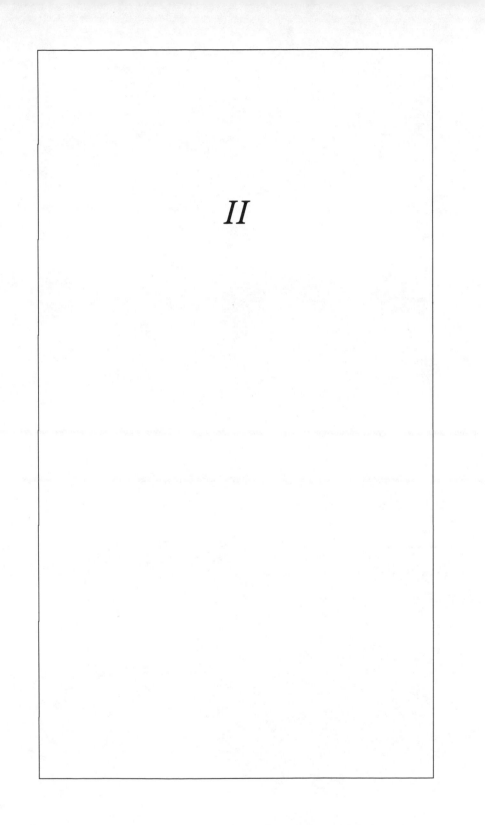

II

In Time of
Plague

Up until just a few years ago, a tour of the exhibition of
artworks on display at the Museum of Natural History [in January
1988] would have left the impression, in most minds, of events
very remote in time, pieces of very ancient history, a strange
and disturbing world now well behind us. For the modern mind,
especially the everyday modern mind now so adept at dismissing
the memories of the great wars of this century with all those
deaths, the notion that great numbers of human beings can die
all at once from a single cause is as far away and alien as the
pyramids. Even more strange is the notion that human death
could ever be so visible, so out in the open. Our idea about
death is that it only takes place privately, in the dark, away from
other people. There is something faintly indecent, wrong, about
dying in full view of the public.

This is understandable, at least for those of us who live in the
Western, industrialized world and have grown to adulthood in
the last half of this century. Dying is now the exceptional thing
to do, almost an aberration, in our culture. We concede the
possibility, even in our bad moments the inevitability, but never
before its time. Moreover, that time, the appropriate and accept-

able moment, is being pushed further and further into the distance ahead. When the century began, the average life expectancy for Americans and Europeans was around forty-six years of age; now the life span for a great many of us to bet on will be nearly double that number. Excluding war, of course, in which case all bets are off.

Indeed, the nearest equivalent to the plagues that afflicted human populations in earlier centuries has become, for our own kind of society, dying itself. We know in our bones that we will all die, sooner or later, but we find it harder and harder to put up with the idea; we want the later to be later still. We think of dying as though it were failure, humiliation, losing a game in which there ought to be winners. Many of us take the view that it can indeed be put off, stalled off anyway, by changing the way we live. "Life-style" moved into the language by way of its connection with getting sick and dying. We jog, skip, attend aerobics classes, eat certain "food groups" as though food itself was a new kind of medicine, even try to change our thoughts to make our cells sit up and behave like healthy cells; meditation is taken as medication by some of the most ardent meditators. We do these things not so much to keep fit, which is a healthy exercise and good for the mind, but to fend off dying, which is an effort not so good for the mind, maybe in the long run bad for the mind.

A certain worry about death is, of course, nothing new; it is the oldest of normal emotions. But it does seem to me strange that the anxiety has acquired more urgency, and plagued us more in the last years of this peculiar century than ever before, during the very years when most of us have had a good shot at living longer lives than any previous human population.

Partly, I suppose, we fear death more acutely because of this very fact. It never occurred to us, until quite recently, that we had any say at all in the matter; death just came. But now, when it seems that we can put it off, for at least as long as we have in recent decades, why not still longer? If instead of surviving for an average lifetime of three or four decades, as used to be the rule, we can stretch it out to eight or nine, why not fix things so that we keep running for twelve or fifteen, and even then why stop?

But now, at the moment of such high expectations, we are being brought up short, with a glimpse—a brief and early one, to be sure—of what living was like long centuries ago. Just within the past decade we are being confronted again by the prospect of mortality on a very big scale. And not the dying that used to nag at our minds, the slipping of our fingerhold on life because of the weakness of old age. This time it is the death of our young adults, including in one group some of the most talented and productive, in another group a great many young people born and raised in the deprivation of our most benighted neighborhoods. Moreover, it is already a near certainty that what we are seeing now is only the beginning of something far more menacing: the transition of an epidemic now localized within a minority of the population into a pandemic affecting everyone. We have only to look at the course of events in parts of Africa, where whole settlements of the inhabitants are now infected by AIDS. The virus is on the loose in Africa, and there is no reason to hope that it will not spread into the community at large in every other part of the world.

The only point I wish to make is that AIDS is, first and last, a problem and a challenge for science. We simply do not know enough about this extraordinary virus—or, as it is already beginning to appear, this set of extraordinary viruses, all closely related but differing in subtle ways—and we have a great deal to learn. And it seems to me self-evident that the only sure way out of the dilemma must be by research. This is not to say that education and behavioral change will not be valuable in the short run ahead as ways to limit the spread for the time being, to slow down the pandemic for a while. Obviously, we should be instructing all young people in what the virus is and what we already know about how its contagion works, and surely we should be trying whatever we can, including more methadone clinics and the free distribution of sterile needles within the heroin-addicted community. But these are not the answers for the long run. If we are to avert what otherwise lies somewhere ahead, we will have to find out how to kill this virus without killing the cells in which it is lodged, or how to immunize the entire population against the virus, or both. These are scientific problems, very difficult and complex, perhaps the hardest ones ever to confront biomedical science. But at the same time they are nothing like blank mysteries; there has already occurred an exceedingly rapid and encouraging progress in the relatively few laboratories now working on the problem, and it is as close to a certainty as anything I can think of in medicine that the AIDS problem can be solved. The work is only at its beginning, in its early stages, and there is a great deal still to be learned.

It comes as something of a surprise, even a shock, to realize that we are faced by a brand-new infectious disease about which we understand so little and can therefore *do* so little. Modern medicine has left in the public mind the conviction that we know almost everything about everything. This is as good a time as any to amend the impression. There are, here and there, pieces of evidence that biomedical science—as we like to call the enterprise, thus combining the now-hardening science of biology with the prestige of medicine—good as it is, is perhaps not entitled to all the credit it gets from the general public.

It needs a glance back to see what happened to improve our longevity. And then, perhaps, a quick and speculative look ahead.

Science had a hand in our betterment, to be sure, and medicine played a modest hand of its own, but nothing like the decisive role that is assumed in some quarters. To read the papers, you might think that medicine turned itself into a full-fledged science all on its own just in the past century, and that is why we now live longer than we did in the eighteenth or nineteenth centuries. This is an agreeable thought for the minds of doctors, but hard to document. It is true that the basic biomedical science in microbiology and immunology, starting a little over one hundred years ago, eventually brought along the applied sciences of immunization and antibiotic therapy. In its new capacity to prevent or cure the major infectious diseases of human beings in Western societies, medicine could now lay a fair claim to being scientific, but only in part. And even here, in what everyone would acknowledge to be the greatest therapeutic triumph of modern medicine, questions about the actual role of science can be raised. Typhoid fever and cholera had already become rare, even exotic diseases in most parts of the industrialized world long before the emergence of antibiotics, thanks mostly to better sanitation, good plumbing, improved nutrition, and less crowded living quarters—in short, a higher standard of living. The morbidity and mortality from tuberculosis was well on its way down, long before the discovery of streptomycin, INH, rifampin, and the rest had opened the way to rational treatment, and the accomplishment was the result of a combination of old-fashioned public health measures plus a spontaneous decrease in susceptibility to TB infection among a better-fed, sturdier population. Rheumatic fever and valvular heart disease had already begun to decline before the introduction of penicillin prophylaxis.

One of the most spectacular triumphs was, of course, the virtual elimination of tertiary syphilis. I have lately been asking around among my neurological colleagues in the New York and Boston academic medical centers, and have found no one who has seen a patient with general paresis in more than a decade. This malignant disorder of the mind, which filled more state hospital beds than schizophrenia when I was a medical student, has very nearly vanished. But how did this happen? Was it the result of meticulous case-finding and early penicillin treatment of all cases of primary and secondary syphilis (of which there is still an abundance in all our cities)? I rather doubt it, especially considering the debilitated condition of most of today's municipal and county health departments. Then how to explain it? With early syphilis still a common, everyday disease, much of it subsiding into latent syphilis without treatment as always happened in the past, why are we not seeing tertiary syphilis, especially syphilis of the brain?

My theory is that it happened because of science, but a misapplication of science. Since it first came on the medical marketplace, penicillin has been scandalously overprescribed and overused. Most patients with upper respiratory tract infections or unaccounted-for febrile illnesses receive penicillin at one time or another, probably in doses sufficient to eliminate spirochetes wherever they are. In the virtual aerosol of penicillin that has affected whole nations since World War II, tertiary syphilis has, quite by accident, been almost eliminated.

Something like this perhaps also accounts for the increasing rarity of rheumatic heart disease in recent decades. The group A beta-hemolytic streptococcus is still among us, but its capacity to launch the old epidemics of throat infections among our school-children may now be sharply restricted by the commonplace use of penicillin for the wrong reasons. If so, it is hard to call this science, but never mind, it works anyway.

The incidence of fatal coronary thrombosis in the United States has changed dramatically, and for the better, since the 1950s, dropping year by year to an aggregate decrease of around 20 percent. No one seems to know why this happened, which is not in itself surprising since no one really knows for sure what the underlying mechanism responsible for coronary disease is. In the circumstance, everyone is free to provide one's own theory, depending on one's opinion about the mechanism. You can claim, if you like, that there has been enough reduction in the dietary intake of saturated fat, nationwide, to make the difference, al-though I think you would have a hard time proving it. Or, if you prefer, you can attribute it to the epidemic of jogging, but I think that the fall in incidence of coronary disease had already begun before jogging became a national mania. Or, if you are among those who believe that an excessively competitive personality strongly influences the incidence of this disease, you might assert that more of us have become calmer and less combative since the 1950s, and I suppose you could back up the claim by simply listing the huge numbers of new mental health professionals, counselors, life-style authorities, and various health gurus who have appeared in our midst in recent years. But I doubt it, just as I doubt their effectiveness in calming us all down.

My own theory is that the 20 percent drop in American coro-nary disease was the result of commercial television, which ap-peared in the early 1950s and has made a substantial part of its living ever since through the incessant advertising, all day and all night, of household remedies for headache and back pain, all containing aspirin. One plausible result of this may have been the maintenance, down the years, of a national level of blood salicylate set at an optimal range for the inhibition of prostaglandin synthetase and, as a result, a 20 percent reduction in platelet stickiness. If I am right, we might predict an upturn in the incidence now that aspirin is having a rather bad press and variants of Tylenol are in fashion.

My contention is that we do indeed have some science in the practice of medicine, but not anything like enough, and we have a great distance to go. Indeed, most of what is regarded as high science in medicine is actually a set of technologies for diagnostic precision—the CAT scan, NMR, and many exquisite refinements of our methods for detecting biochemical abnormalities of one sort or another. But these have not yet been matched by any comparable transformations in therapy. We are still confronted by the chronic disablements of an aging population, lacking any clear understanding of the mechanisms of these diseases—dementia, for instance, or diabetes or cirrhosis or arthritis or stroke and all the rest on a long list—and without knowing the underlying mechanisms we lack new therapeutic approaches. To be sure, we do have some spectacular surgical achievements in the headlines—the transplantation of hearts, kidneys, livers, and the like—but these are what I have called halfway technologies, brought in to shore things up after the still-unexplained diseases of these organs have run their course. And these measures, plus the new advances in diagnostic precision, account for a large part of the escalating costs of health care today. It seems obvious, to me anyway, that the only practical policy for bringing down those costs will be by more and more basic research in biomedical science, in the hope and expectation that we can then begin to understand, at a deep level, the underlying events in human disease. Sooner or later I am confident that this will be accomplished, and I hope for the sooner. This is the mission, by the way, of the National Institutes of Health, and I hope nobody tries to disturb, for any political purpose, the functioning of this extraordinary institution, either by "privatizing" it, as some ideo-

logues have openly proposed, or by micromanaging its affairs, as
congressional committees are sometimes tempted to do. I have
thought many times in recent years that if you are looking around
for some piece of solid evidence that government really can work
brilliantly well—and these days many are indeed looking despair-
ingly for such evidence—take a glance at NIH, in its singular
way the finest social invention of the twentieth century.

One reason why medical history is not much taught in medical
schools is that so much of it is an embarrassment. During all the
millennia of medicine's existence as a profession, all the way
back to its origins in shamanism, public expectations have been
high, demanding, and difficult to meet. First among the demands
was to recognize the existence of a disease, explain it, and then
make it go away. For most of its history, medicine has been
unable to do any of these things, and its sole advantage to a
patient seeking help lay in the doctor's capacity to provide reassurance and then to stand by while the illness, whatever, ran its
course. The doctor's real role during all those centuries was
more like that of a professional friend, someone at the bedside,
standing by. When he was called a therapist, it was more in the
etymological sense of the original Greek *therapon*: Patroclus was
therapon to Achilles, part servant, part friend; he stood by him
in his troubles, listened to his complaints, advised him when he
could, put up with him, finally even died for him. Therapy was
a much more powerful word for the ancient Greeks than it is
today, and somewhere along the line it was picked up by the
doctors. In its ancient meaning, it carries inside all the obligations
that medicine and the other health professions have for patients
with AIDS: to stand by, to do whatever can be done, to comfort,
and to run all the risks that doctors have always run in times of
plague.

When I arrived at Harvard Medical School in 1933, nobody
talked about therapeutics as though it were a coherent medical
discipline, in the sense that pharmacology is today. To be sure,
there were a few things to learn about: digitalis for heart failure,
insulin, liver extract for pernicious anemia, vitamin B for pellagra,
a few others. By and large, we were instructed not to meddle.

Our task was to learn all that was known about the natural history of disease so that we could make an accurate diagnosis, and a reasonably probabilistic prognosis. That done, our function as doctors would be to enlist the best possible nursing care, explain matters to the patient and family, and to stand by.

Moreover, we were taught that this was not only the very best kind of medicine—not to meddle—it was the way medicine would be for the rest of our lives. None of us, certainly not any of the medical students, had the faintest notion that our profession would ever be any different from what it was in the 1930s. We were totally unprepared for the upheaval that came with the sulfonamides, and then with penicillin. We could hardly believe our eyes on seeing that bacteria could be killed off without at the same time killing the patient. It was not just an amazement; it was a revolution.

That made two revolutions. The first, that the medical treatments in common use for all the centuries before our time didn't really work, did more harm than good, and had to be given up, left us with almost nothing to do to alter the course or outcome of a human illness. And then the second, that a form of treatment based on fundamental research, tested in experimental animals, then tested in controlled experiments in human beings, was possible. Medicine, it seemed, was off and running. Nothing seemed beyond reach. If we could cure streptococcal septicemia, epidemic meningitis, subacute bacterial endocarditis, tuberculosis, tertiary syphilis, typhoid fever, even typhus fever, was there anything we could not do? Easy one hand, we thought.

By the 1950s it was beginning to become clear that solving all the rest of medicine's problems was going to be a lot harder than we had thought. Some of us had overlooked the fact that the conquest of bacterial infection and the success in immunizing against certain viral diseases had not simply fallen into the lap of medicine. These things could not have happened at all without the most penetrating insight into the underlying mechanisms of the diseases we were able to treat or prevent so effectively. Not to say that we had much comprehension of how a meningococcus goes about producing meningitis, or how a pneumococcus does its special work in the human lung. These were blank mysteries, and to some extent the mechanisms of pathogenesis remain mysteries today. But what we did know were bits of crucial information: the names and shapes and some of the metabolic habits of the microorganisms involved, the specific pathologic changes they produced in human tissues, and the ways in which they spread through human communities. This information had not dropped into our laps. It came as the outcome of decades of what we would be calling basic biological science if it were going on today. Without the research in microbiology and immunology done in those decades, we would never have entered that first period of modern therapeutic medicine.

Now things are beginning to change. A problem like cancer, which seemed simply unapproachable in the early 1970s, too profound even to make good guesses about, has turned suddenly into the liveliest, most exciting and competitive, and among the most allusive puzzles in all of biology. Instead of appearing to be one hundred different diseases, each requiring eventually its own individual solution, it now seems more likely that a single, centrally placed genetic anomaly will be found as the cause of all human cancers. The new information, coming in as cascades of surprise from laboratories all round the world, will almost surely lead to novel therapeutic and preventive approaches that cannot be predicted at the present time.

There is something of the same excitement and anticipation within the rapidly expanding community of neurobiologists, keeping in the same close touch with each other across all international boundaries. The brain has been transformed within just the past few years from a computerlike instrument of unimaginable complexity to a system governed by chemical messages and run by the specific receptors of scores, more likely hundreds, of short peptide chains of command.

Much of what is now happening in both cancer and brain research is the outcome of basic research that had neither of these problems in mind when the work was started, even when the definitive aspects of the work were well under way. The new field of molecular genetics began to catch glimpses of its potential when the restriction enzymes were discovered, and once the technology of recombinant DNA had evolved to the point where specific molecular probes were at hand, it became clear that the most powerful methods for studying the deep functions of cancer cells were also at hand. Meanwhile, the new technique of cell fusion had led to the development of hybridomas, and monoclonal antibodies had become indispensable for the study of gene products and cell surface antigens. But the people whose work made these techniques possible were not, at the outset of their work, aware that they were putting together a totally new set of approaches to the cancer problem. Early on, nobody engaged in the work could have predicted how it would turn out or where it would lead, nor can they predict the future benefits today, only that the work is engrossing and fascinating, and filled with surprises yet to come.

This is the way it will go for the rest of medicine, if fundamental biomedical research continues at its present pace and scope. The dementias of the aged, coronary artery disease, arthritis, stroke, schizophrenia and the manic-depressive psychoses, chronic nephritis, cirrhosis, pulmonary fibrosis, multiple sclerosis—any disease you care to name—will be opened up for closer scrutiny by discoveries still unpredictable in biology itself. The information needed for the future cannot be planned by committees or commissions, and there is no way for bureaucracies to decide which pieces of information are needed first, or at what time. In basic science the shots can never be called in advance—or by definition the enterprise is not basic science.

This sounds an optimistic way of looking at the future of medicine, for what I am predicting is that no human disease is any longer so strange a mystery that its underlying mechanism cannot be understood, or got at. There are some who will quarrel with this view of things, arguing that I am oversimplifying matters of great complexity. They will say, they do say, that the idea of single causes for disease is outdated and wrong, that people who talk about single causes are influenced too much by the infectious diseases. Today's chronic ailments, they will assert, especially the chronic diseases of aging, do not have single causes, they are multifactorial, whole systems of things gone wrong. The environment itself is what needs fixing, along with life-style, diet, exercise, and a new kind of human personality to boot.

Maybe so, but I doubt it. I do not doubt that a good many influences affect the incidence and severity of human disease. Pneumococcal pneumonia is quite a different disease when it occurs in a chronic alcoholic, or when it afflicts a very old person, or a person with immunodeficiency. But the pneumococcus is still at the center of things, the chairman of the board. I think there will be a chairman of the board, perhaps at the head of a whole committee of other mechanisms in the senile dementias, or in schizophrenia, or cirrhosis, or all the rest. And once identified, he can be got at, just as the spirochete of syphilis, running what seems to me the most complicated, multi-organ, multi-tissue, multi-mechanism of all the human diseases I have ever learned about, can be got at. Once got at, the simple act of lifting out the spirochete results in the switching off of all the other events. There is no good explanation for this, but it is, I believe, the way things work in human disease.

It is this approach, now getting under way but needing much more expansion, that is needed for the AIDS problem. Through the efforts of NIH, the Pasteur Institute, and a few other laboratories here and abroad, an intelligent investment is already being made—still on a modest scale considering the necessity—in scientific research on the biology of AIDS and its virus, HIV. Considering that the virus has been recognized for less than a decade and that it is one of the most complex and baffling organisms on earth, the progress that has already been made in the laboratories working on it is an astonishment. I have never observed, in a long lifetime looking at biomedical research, anything to touch it. If this disease had first appeared well before the research technologies of molecular biology had developed the marvelous tool of recombinant DNA, we would still be completely stuck, unable even to make intelligent guesses about the cause of AIDS. Thanks to these new methods, which emerged from entirely basic research having nothing at all to do with any medical problem, we now know more about the structure, molecular composition, behavior, and target cells than we know about any other virus in the world. The work, in short, is going beautifully. But it is still in its early stages, and there is an unknown distance still to go. At the moment, there seem to be three lines of research holding the most promise, and there is a conspicuous shortfall already in the funds needed for each of these lines.

One approach, the most direct of the three but also the most difficult and unpredictable, is in the field of pharmacology. We need a new class of antiviral drugs, capable of killing off the virus inside all the invaded cells without killing the cells themselves. These drugs must be comparable in their effectiveness to the antibiotics which came into medicine for use against bacterial infections in the 1940s. We already have a few partially active drugs; these may turn out to be the primitive precursors of such a class—AZT, for example—but their effectiveness is still incomplete, temporarily palliative at best, and their toxicity is unacceptable. However, there are no theoretical reasons against the development of decisively effective antivirals, including drugs to stop the replication of retroviruses like HIV. What is urgently needed, indispensable, in fact, is some new and very deep information about the intimate details of retroviruses and the enzyme systems that enable them to penetrate and multiply within the target cells that are their specialty in life—in short, more basic research of the most fundamental kind.

Second, we need an abundance of new information about the human immune system. The only imaginable way to prevent the continuing spread of HIV—even when and if we have in hand an antiviral drug that really works to control infection in individual cases—is to produce a vaccine. What this means is that more information is needed concerning the molecular labels at the surface of the virus, and which among these labels represents a point of vulnerability for an immune response. Since it is already known that this particular virus has the strange property of changing its own labels from time to time, even the labeling of the same HIV virus isolated at different stages of the disease in the same patient, it will be no easy task to produce a vaccine.

A small number of vaccine trials are already under way in small cohorts of human subjects. There is no reason to be optimistic about these at the present time, nor is there any feasible way to hurry things up. With luck, a lot of luck, some laboratory may succeed in identifying for sure a stable and genuinely vulnerable target molecule in HIV, and then a vaccine will be feasible. But as things stand now, there can be no assurance that a vaccine prepared against this year's crop of HIV viruses will be any more effective against the viruses at large five years hence than that an influenza vaccine prepared several years ago will be of any use against next winter's outbreak of flu.

A third line of research involves the human immune system itself, the primary victim of the AIDS virus. Actually, most if not all patients with AIDS die from other kinds of infection, not because of any direct, lethal action of the virus itself. The process is a subtle one, more like endgame in chess. What the virus does, selectively and with exquisite precision, is to take out the population of lymphocytes that carry responsibilities for defending the body against all sorts of microbes in the world outside, most of which are harmless to normal humans. In a sense, the patients are not dying because of HIV, they are being killed by great numbers of other bacteria and viruses that can now swarm into a defenseless host. Research is needed to gain a deeper understanding of the biology of the immune cells, in the hope of preserving them or replacing them by transplantation of normal immune cells. This may be necessary even if we are successful in finding drugs to destroy the virus itself; by the time this has been accomplished in some patients, it may be that the immune system has already been wiped out, and the only open course will be to replace these cells.

This had already become one of the liveliest fields in basic immunology long before the appearance of AIDS, and what is now needed is an intensification of the research. In my own view (perhaps biased because of my own background in immunology), it is the most urgent and promising of all current approaches to the AIDS problem.

To sum up, AIDS is first and last a scientific research matter, only to be solved by basic investigation in good laboratories. The research that has been done in the last few years has been elegant and highly productive, with results that tell us one sure thing: it is a soluble problem even though an especially complex and difficult one. No one can predict, at this stage, how it will turn out or where the really decisive answers to our questions will be found, but the possibilities are abundant and the prospects are bright. It is particularly encouraging that the basic research most needed is being conducted by collaborative groups in both academic and industrial establishments. This is a new phenomenon in this country, well worth noting in the present context. Up until just recently, the past decade or so, the university laboratories and their counterparts in the pharmaceutical industry tended to hold apart from each other, indeed rather looked down their noses at each other. It took the biological revolution of the 1970s, and specifically the new technologies of recombinant DNA and monoclonal antibodies, to bring the scientists from both communities into a close intellectual relationship, with each side now recognizing that the other could offer invaluable contributions to research on ways to intervene in the mechanisms of human disease. And now, as the science moves along from one surprise to the next, especially in the fields of molecular biology and virology, the lines that we used to think of as separating basic and applied research into two distinct categories have become more and more blurred. The academic and industry scientists recognize that they are really in the same line of work, and research partnerships of a new kind are being set up all over the place; the scientists, by and large, are moving as fast as they can. It is already clear that there is not enough talent to go around. However, I take this to be an exceedingly healthy transformation in our institutions. The response must be the recruiting and training of more bright young people for the work ahead.

I should say a few words here about the possibilities for spin-offs from AIDS research. Briefly, they are endless. The immunologic defect produced by HIV brings about an increased vulnerability to Kaposi's sarcoma, and the same class of immune cells may play an important role in the defense of humans against many other, perhaps all, types of cancer. The dementias that occur in terminal AIDS infection resemble other types of human dementia, and new information about the former should shed light on the latter. If we can learn how to block the HIV retroviruses without killing the cells in which they are lodged, we will find ourselves armed with a new, more general class of antiviral drugs, of very general usefulness. It used to be said as a truism in medical school, long ago when I was a medical student, that if we could ever reach a full understanding of the events that occurred in the human body as the result of one single disease, syphilis, we would know everything in the world about medicine. This is the hunch I have today about AIDS.

It is obvious that the science that is needed for the AIDS problem in the years just ahead will only be possible with the support of large expenditures of public money. The commitment will demand public support, obviously. But I do wish, and fervently hope, that it can somehow be taken out of politics itself. AIDS is not, emphatically not, a political problem. Questions about testing people for the virus, and whom to test, even how to test, and how to preserve confidentiality, are proper questions for the public health professionals who have known for years how to do this kind of thing, and they are not proper questions for politicians running for office. AIDS is a complicated scientific puzzle, as complex as any yet faced by biomedical science, and more urgent than any other I can think of. It is, in short, an emergency for science, the best science to be done by the best scientists we can assemble, in as international an effort as we can bring into being, and as quickly as possible.

AIDS and
Drug Abuse

*T*here are enormous problems confronting the health of modern society, worldwide. They are linked together as health problems only accidentally and in a very restrictive sense: AIDS is being transmitted among some sections of the population, especially in this country, by the contaminated needles of intravenous heroin users. That, so far as we know, is the only connection between the problem of AIDS and the problem of drug addiction, but perhaps it is enough of a connection to encourage us to look more deeply into the pathology of society for other points of linkage.

Taken together, these two disasters—AIDS and drug abuse—will soon be draining away more of the country's resources than any other health problem, perhaps more than all our other health problems combined.

By itself, within just the next few years, AIDS may, and probably will, be costing whatever the sums of money are that are needed to care for upwards of one million, mostly young people, who will require the highest and most expensive technologies available to medicine, for prolonged periods, months, even years, of slow, painful, and (as things now stand) absolutely inevitable

dying. I leave it to others to make guesses at the economics of
health care and to set dollar figures on those costs which lie just
ahead of us. I cannot even guess at the magnitude of dead loss
to the country, in purely financial terms, of such enormous numbers of our youngest citizens. I do know that many of them will
be extremely intelligent people of high talent and potential within
the homosexual community, and I am quite sure that the time
has already come for the disease to begin its spread into our
heterosexual population as well. At an uneducated guess, I would
put the total cost in the very high scores of billions of dollars,
and rising year after year.

As to drug addiction, our other national catastrophe (nothing
less), the dollar costs are already insupportable, quite aside from
the costs in lives and productivity. Incidentally, where did we
ever find that evasive term "substance abuse" for such a problem? It always sounds to me like words being put to use in order
to hide something from our minds. In plainer language, the big
problems are heroin, cocaine, and crack, relatively new puzzles
for our kind of society, and alcohol, an ancient and even more
intractable problem. These are quite separate dilemmas, not to
be solved by any approach that tends to lump them together.
To be sure, I suppose they all share origins of some sort in
defects in the moral fiber of people, whatever that may mean,
and they exist and can fairly be viewed as evidences of slippage
in the whole society. Something has indeed gone wrong, and the
cost of that something, whatever it is, cannot of course be measured only in dollars. But the dollar costs are there, sure enough,
and need more than my competence to calculate them. Not just
the illnesses and deaths—again, affecting mostly our youngest
citizens—but add the outlays for the national constabulary just
for coping with the drug trade, plus the border controls, the
criminal justice system, the correctional institutions, and all the
rest. There, by the way, went another evasive term unaccount-
ably in current usage: "correctional." Prisons, you mean, and
more prisons to build every year, just for the drug trade. And
not much else to offer.

Or do we really have, somewhere, in adequate numbers and

operating effectively, institutions for the correction of heroin and cocaine addiction, or, for that matter, even alcoholism? Well, we do have a few methadone clinics here and there for the heroin addicts in some of our cities, nothing like enough to meet the need, underfinanced and understaffed. And we have, here and there, a very small set of places for drying out the alcoholics of well-off families.

I have not yet heard of any organized effort, on any realistic scale, designed to inquire into whatever the things are that have gone so wrong in our society as to induce so many of our youngest people to attempt escaping from precisely that society by way of drugs (not to mention suicide). And if there were the possibility of carrying out such an inquiry and arriving at some solutions, I am convinced that it would cost a vast sum of money and I doubt very much that the money would be made available, given our society's present attitude toward itself.

One thing is certain: none of the money generated by the triumphant heroin and cocaine industries, not one penny, will ever be ploughed back into the nation's economy, in taxes or in philanthropy, to be used for the rescue of children. To the contrary, enterprises as profitable as these transnational businesses can be relied on to invest as much of their funds as possible into extending their markets, bringing in more customers every year. Talk about growth industries! Just think back: twenty-five years ago it seemed a relatively small, manageable problem, a nasty one to be sure but confined to the minority community living in the poorest sections of our cities, a few billion dollars' annual take for the heroin merchandisers, marijuana a relatively benign drug mostly used, we thought, by a handful of black musicians, cocaine an exotic substance not much talked about in public and rarely encountered beyond the ghettos. And now look: cocaine has spread from the outcast zones of our cities, and is now in fashion, high-style in fact, in midtown, even down on the Wall Street sidewalks. I suspect that the combined net profits being generated each year by the sagacious people running heroin, cocaine, and marijuana into this country must exceed the total incomes of the breakfast food and tobacco industries, plus a large piece of the liquor business, plus what's left of television sets and steel. But although it is a dominant part of the American economy, the dollars flow south and overseas. By the way, perhaps someone can tell me whether any of those dollars show up in our annual trade deficits. If so, can anyone forecast what it is likely to be five years hence?

So much for the economic side of the problems, and so much does seem a tremendous lot even if I cannot do the numbers. We are in for something very big indeed, unless we can learn quickly how to mend our ways. I must confess that I am at this time not optimistic about any magical property of money as a fix, not anyway until we have learned a lot more than we know at present. For a country already beset by a range of economic dilemmas beyond the grasp of most of us, we would do well to become more frightened than most of us seem to be. The economic impact of AIDS and drug addiction, added on top of all our other problems, could damage the culture beyond repair.

The academic economists must be finding these matters an intellectual challenge, but I confess to total ignorance as to what they may be up to. I hope nobody plans to ask me what the medical profession, or any of the sciences connected to medicine, or the psychiatrists and psychologists or the other social scientists, are doing about the drug problem. If asked, my glum answer would have to be not much. Apart from methadone maintenance—a brilliant idea when first thought up by Vincent Dole and Marie Nyswander in the 1960s, but now almost an irrelevancy because of inadequate public support in the early years and because now the heroin problem is so far outclassed by the crack cocaine problem—the professionals have very little to brag about. We use the term "counseling" all over town, with the unwarranted implication that simply talking to young addicts, or listening to them, had some effect, which I greatly doubt. And now, the term in high fashion is "education." I shall have more to say about education in a moment.

On the brighter side, if you can use such a word, is the intelligent investment that is already being made—still on a modest scale considering the necessity—in scientific research on the biology of AIDS and HIV. Considering that this virus turned up only a few years ago, and that it is one of the most complex and baffling organisms on earth, the progress that has already been made in the laboratories working on it is an astonishment. AIDS, therefore, is essentially a matter for scientific investigation and its solution can come only from basic research in good laboratories.

Now I would like to say a few things about education and the AIDS problem, and about education and the drug addiction problem. Although I dislike using a word like "education" for the process, I do agree that the only way to intervene *today* in the spread of AIDS through the population is by changing the behavior of individuals. Education seems to me an inappropriate word, implying as it does the learning of a lot of things about the world. The best information that is available to be conveyed these days concerns only a very short list of very big things, involving the limiting of sexual contacts, the use of condoms, and possibly abstinence (I can think of no others). Whether this short list is called education or counseling or whatever, it is obviously useful advice, and it may be effective to some extent in restraining the spread of the virus, but I doubt very much that it will stop the epidemic. The virus is already too far afield, affecting both sexes, and it is simply bound to spread. Nevertheless, I am, with these reservations, on the side of those calling for an urgent effort, within all our schools and involving all the media, to make the facts of AIDS as plainly and candidly clear as possible to everyone, including as many of our preadolescent children as can be reached. But, I should reiterate, the virus is already out in the open and will continue to spread until we solve the science problems.

But as to the drug addiction problem, I do not feel even this degree of modified optimism about education, not in the way the word "education" is being so widely used these days, and certainly not for any science I can perceive on the horizon.

It seems to me that the spread of drug addiction, which had

its start among the young people in the most impoverished sections of our population, now involving as it does even young children, signifies that something has gone so badly wrong with the way we live together, and the way children are treated in our society, that we should begin thinking about education in an entirely new light.

I do not propose here to discuss, beyond mentioning it, the slow destruction by neglect and default of public education at the primary and secondary school levels. Everyone knows about this, everyone talks about it, everyone knows that the public schools in every large city have been allowed to decay over the past generation into custodial institutions for the children of poor people, and the private and parochial schools for more affluent white children are only marginally better places for learning how the world works. I am glad to hear that some of the nation's corporate leaders have become concerned about the problem and its obvious implications for the quality of the present and future American work force. I believe this is a matter where intervention and support from the private sector can pay, in the long run, high dividends. As a society, we are so obviously guilty of neglecting the minds of our school-age children that I need only say that this is so, and move on.

The educational neglect that I am more worried about exists at a deeper level and will, in the long run, cause much worse damage to our culture if it is not corrected. It is the neglect, amounting to abuse, of the minds of very small children, throughout the years of early childhood and long before kindergarten or first grade. The children most affected are, of course, black and Hispanic, the poorest of our cities' poor, many of them in single-parent households, an increasing number in recent years with no parents, no family at all, housed like zoo creatures for months in the pediatric wards of city hospitals. Hundreds of thousands of our children are receiving this kind of start in life each year, and fifteen or twenty years from now we will be wondering, as we wonder about their predecessors today, and about the literally hundreds of thousands of homeless children lost in our streets, how it was that so many of our fellow citizens turned out so badly, so antisocial, so criminal.

We seem to have forgotten, or never learned, what young children really are, and how special their minds are. Most of us tend to think of early childhood as a primitive stage of life, a sort of deficiency in the mind that will, with time, be outgrown. What we keep overlooking is the sheer tremendous power, unique in the brain of a young child, never to be matched again later in life, for *learning*. A child three or four years old can become proficient in three or four different languages at the same time, under the right circumstances, when living at close quarters with children of other nationalities. The very small children of Turkish migrant families living in Germany spend their evenings trying to teach their parents how to speak good German.

Language acquisition is a special gift of children, we say, but we say this as though it were the only gift, as though receptors for language were somehow wired into the brain of a young child and then replaced in later years by something else, more useful and more adult. I do not believe this. I am persuaded, by things written by some of the specialists who have spent their lives studying young children, and also by observations of my own, that young children possess minds that are fabulously skilled at all sorts of feats beyond mere language. They have receptors wired in for receiving the whole world; they are biologically specialized for learning. But if they are deprived, in their earliest years, of an environment in which learning can be stimulated, they may pass through those years and emerge with quite different minds. Not stupid or mindless, mind you, just different, with a skewed concept of the world and its arrangements.

The experience that is above all others in its importance for the modeling of a young child's mind is, in my view, a combination of affection and respect. This magical formula is provided by parents, especially by mothers, and when it is lacking, or withdrawn, society for its own future sake must somehow replace it in the young child's environment. There are some private agencies concerned with the surrounds of preschool children, and a few publicly sponsored efforts—Project Head Start, for one— have been tried out with remarkable success.

It seems strange to me that the older I get, the more obsessed I become with this particular aspect of the nation's huge

educational problem. Somehow I have come to believe that unless we can discover how to bring that combination of affection
and respect into the daily and nighttime lives of our children—
most especially our poorest children—everything else we hope
to do for the improvement of education later in their lives will
be one dead end after another. We need reforms all over the
face of education, up to and including our universities, of course.
But if we hope for a generation of young people who will not be
casting around nervously for ways to escape the world, including
cocaine and heroin and hashish and the rest, we should be setting
our highest priority on the education of preschool children, and
we should be starting the effort in those wretched demeaning
parts of our cities that we dismissively call the ghettos. I shall
leave it there.

The Odds on
Normal Aging

*I*n spite of today's ignorance about so many different diseases, including most of the chronic illnesses associated with aging, there is the most surprising optimism, amounting to something almost like exhilaration, within the community of basic biomedical researchers. There has never, I think, been a time quite like the present. Most of the investigators, the young ones especially, have only a remote idea of the connection of their work to human disease problems, although they have an awareness that sooner or later something practical and useful may come from their research. But this possibility is not the driving force behind their endeavors. The main impetus is that researchers are becoming confident about finding out how things work. This is true for the immunologists threading their way through the unimaginably complex network of cells and intercellular messages that make up the human immune system. It is true for the experimental pathologists and biochemists at work on the components of the inflammatory reaction who turn up new regulatory cell products and signaling devices almost every month. The cancer biologists are totally confident that they are getting close to the molecular intimacies of cellular transformation, and the virologists are riding high. Out in the front lines are the molecular biologists and the geneticists in possession of research techniques that permit them to ask (and answer) almost any question that pops into their minds.

The science underlying the aging problem, including the problem of cancer, relies on basic research for finding out how things work in a normal cell; but, at the same time, it is a venture in applied science, for no one doubts that we will not only be provided with a clear comprehension of how cells age and become cancerous, but we will very likely develop some useful methods to reverse or control these conditions. Aging research provides a good example of the way in which biomedical science in recent years has become an international venture moving back and forth across national boundaries. If the problems of human aging and of cancer are solved by one or more of today's research approaches, the ultimate solution cannot be fairly claimed by any nation or by any particular laboratory or group of laboratories. The work has reached its present state of high promise as the result of an intricate network of international collaborators, and it will have to progress in the same way if it is to be ultimately successful. To be sure, there will be the usual strident claims on priority by whoever is successful in putting the final piece of the puzzle in place, but everyone will know, and I hope remember, that the puzzle itself could not have been shaped into being without the most intense international cooperation over at least the last four decades.

Crucial bits of information, indispensable for framing the questions that lie ahead, are coming from laboratories all over the world. The scientists in the field have been keeping in such close touch with each other that everyone knows the contents of the latest paper months before its publication. The results of the latest experiment in Edinburgh or Boston are known to colleagues in Melbourne or Tokyo almost as soon as completed. The mechanism for the international exchange of scientific information is informal and seemingly casual, resembling gossip more closely than any other sort of information system, except that gossip has a reputation for unreliability, and this scientific exchange is generally solid and undistorted. The information is not just passed around automatically; it is literally given away, a curious phenomenon in itself, looking something like altruism in the biological sense of that term. It is intuitively recognized by the participant that free exchange of data is the only way to keep the game going. If one's own new information is withheld from another laboratory in the interests of secrecy, the flow of essential information from that laboratory will itself be stalled, and the whole exchange may slow down and perhaps stop altogether.

There was not this sense of optimism in cancer research not very long ago. In the early 1970s, while I was busy doing research on various problems of immunity and infection, I could not have imagined a scientific problem less attractive than the problem of cancer. I thought of cancer research as an impossible undertaking, and I had the same hunch about aging: they appeared, at that time, insoluble. I wondered at the zeal and courage of my colleagues who were engaged in research in those fields and felt sorry for their entrapment in a scientific "blind alley." Aging and cancer then seemed to be not single problems but a hundred different problems, each requiring its own separate solution, and all of the questions that arose were not only very hard questions but had the look of being unanswerable ones. How could anyone begin to seek answers to questions about a process such as cancer that seemed to encompass almost every discipline in biomedical science virology, immunology, cell biology, and membrane structure and function? Aging as a research problem seemed worse.

I knew that a few people here and there were doing clinical research and had discovered that a few chemical agents had a modulating effect on the leukemias of childhood. The chemicals were dangerous and toxic and difficult to handle, and while the clinicians were hopeful, as an outsider I was not. If any young postdoctoral student or M.D. had asked me about the advisability of going into cancer or aging research in the late 1960s, my advice would have been to stay away and pick a field where things were moving along nicely, like immunology. Even in the early 1970s when the National Cancer Program was being put together for the declared purpose of launching the so-called conquest of cancer, with a substantial infusion of new funds for the support of cancer research, I and many of my colleagues remained skeptical about the whole venture. It is just too early for a crash program, we said. Biological science is not ready for this. We do not know enough. Some of us even said in testimony before various congressional committees that the problem of cancer would not become approachable for another fifty years.

Then, still in the early 1970s, things began to change at a great rate, and they have been changing with stunning speed ever since, astonishing everyone. Now work that was state-of-the-art just a few years ago has already an antique look, and the most talented of the rising generation of scientists are streaming into cancer research everywhere. Best of all, they are entering the field because it is becoming one of the most exciting and enchanting of all problems in biology and alive with possibilities; it is beginning to look like an approachable problem and even a soluble one.

What has happened to bring about this change? I suppose money had something to do with it at the outset, but it was not primarily responsible. What happened was that basic science did what basic science tends to do every once in a while; namely, it produced by luck an overwhelming, totally unplanned-for set of surprises. There were two outstanding and memorable surprises, both of which turned out to be indispensable for research not only on cancer but on a whole range of human diseases, including aging. The first astonishment was the technology of recombinant DNA, enabling an investigator to ask almost any question about the intimate details of a living cell's genes and then to receive sharp, clear answers. Using these techniques, it soon became plain that there were cancer genes and also other genes that restrain the cancer process. Now we have learned how chemical carcinogens and cancer viruses can change and switch on such genes. The whole field of specific intracellular systems and their specific signals and receptors is being transformed by the new recombinant DNA technologies. The other big development was the discovery of cell fusion and then the formation of cell factories for making monoclonal antibodies.

With these tools, it is now possible to identify gene products that are elaborated by cancer cells and other abnormal cells, and to examine with a high degree of precision and specificity the changes that take place in the cell membrane when the normal cell is switched to a cancer cell. Probably we will soon discover why a normal cell becomes an aging cell. The virologists are having a perpetual field day; the immunologists are ready to claim the whole problem of cancer as well as aging as their own; and the biophysicists, the nucleic acid chemists, the geneticists, the cell biologists are falling all over each other in the race to final answers. There has never been a period of such high excitement and exuberance and confidence in any field of biology. It begins to resemble what one reads about in the early days in twentieth-century physics, when quantum theory was just beginning to take shape. Biological science is in the process of upheaval by what is being discovered about cell biology, and nobody can be sure what lies ahead beyond the certainty that there will be brand-new information at the deepest levels and therefore important and useful.

Two aspects of this scientific phenomenon seem to me remarkable in terms of public policy and the implications for the future of the health sciences. First, no committee anywhere could possibly have predicted any of the events that have taken place. We are observing basic science at its best, moving along from one surprise to the next, capitalizing on surprise, following new facts wherever they seem to lead, taking chances and making guesses all the way and driving the problems along toward their ultimate solution; the researchers are not following any rulebook or any long-range plan but playing hunches. The second remarkable feature is the sheer spread of biomedical territory that is opening up as the work goes along. Cancer itself is turning into a soluble problem, although I have no way of guessing at the likeliest outcome. The answer may lie in gaining control pharmacologically or immunologically over the switching mechanisms responsible for the activation and transformation of cells, or it may lie in the chemical nature and mode of action of protein gene products coded by oncogenes. On the other hand, a set of signaling events occurring at the cell surface or within the cell membrane may be responsible for transforming a normal cell into a neoplastic cell. The point is that whatever it is, the research technologies are becoming sufficiently powerful and precise that it is almost unthinkable that the inner mechanism can long remain hidden. And at the same time, cell biology advances as a huge new enterprise in biology, quite independent of the cancer problem. Within a decade, cellular immunology has become one of the most sophisticated of biomedical disciplines, capable of opening the way into problems of the so-called autoimmune diseases, such as rheumatoid arthritis, diabetes, and multiple sclerosis. A combination of the forces of modern virology and cellular immunology has opened up new approaches to the mechanism of damage to pancreatic islet cells in diabetes and its eventual reversal.

Neurobiology has also begun to take off in the last few years. The discovery of the endorphins, followed by a cascade of other internal hormones secreted by brain cells into the brain itself, is turning the central nervous system from an incomprehensible, computerlike, hard-wired apparatus into a chemically governed system of signals. Experiments with primitive marine organisms are revealing neural mechanisms and structures involved in short-term and long-term memory. Selective enzyme deficiencies have been observed in the brain tissue in patients with Alzheimer's disease, while other forms of senile dementia are known to be caused by a so-called slow virus, the CJ agent.

Malignant hypertension has become a treatable disease, although when I was an interne, it was a sure death sentence. Moreover, new drugs have been deliberately designed for inhibiting particular enzymes that lead to hypertension. Cardiovascular pharmacology is emerging as a field in which the chemists can call the shots in advance in making their chemicals. The whole of biomedical science is on the move as never before in the history of medicine.

I do not know what will happen over the next twenty years, but my guess is that we are on the verge of discoveries that will match the best achievements in infectious disease a generation ago. As we develop new decisive technologies that are based on a deep understanding of disease mechanisms, my guess is that they will turn out to be relatively inexpensive compared to the kinds of measures that medicine is currently obliged to rely on. A genuine high medical technology will make an enormous difference to medical practice in the decades ahead, provided that we keep the basic biomedical sciences going and couple them as congenially as possible to clinical research.

It is abundantly clear that the problem of aging is a proper field for scientific study and one of the broadest of all fields in human biology. The array of specific questions to be asked is long and impressive, and each question is a hard one requiring close and attentive scrutiny by the best practitioners of basic science and clinical medicine. And, as the answers come in, there is no doubt that medicine will be able to devise new technologies for coping with the things that go wrong in the process of aging. This is an optimistic appraisal but not overly so, provided we are careful with that phrase "things that go wrong." There is indeed an extensive pathology of aging, one thing after another goes wrong, failure after failure, and the cumulative impact of these failures is what most people have in mind and fear as the image of aging. But behind these ailments, often obscured by the individual pathologies, is a quite different phenomenon: normal aging, which is not a disease at all, but a stage of living that cannot be averted or bypassed except in one totally unsatisfactory way. Nonetheless, we regard aging these days as a sort of slow death with everything going wrong.

I would like to separate these two aspects of the problem, for I think that the former can be approached directly by the usual methods of science. The list of pathologic events is long but finite. Way at the top are the disorders of the brain leading to dementia, which is the single most dreaded disease by all aging people and their families; then there is cancer, of course; bone weakness; fractures; arthritis; incontinence; muscular wasting; Parkinsonism; ischemic heart disease; prostatic hypertrophy; pneumonia; and a generally increased vulnerability to infection. They represent the discrete, sharply identifiable disease states that are superimposed on the natural process of aging, each capable of turning a normal stage of life into chronic illness and incapacity, or into premature death. Medicine and biomedical science can only attack them one by one, dealing with each by the established methods of science, which is to say, by relying on the most detailed and highly reductionist techniques for research. If we succeed in learning enough of the still obscure facts about Alzheimer's disease, we will have a chance to turn it around sooner or later, and in the best of worlds, to prevent it. But lacking these facts, we have no way at all to alleviate it or to help the victims. If the scientists are successful, we can hope for a time when the burden of individual disease states can be lifted from the backs of old people, and they are then left to face nothing but aging itself.

And what then? Will such an achievement remove aging from our agenda of social concerns if old people do not become ill with outright diseases right up to the hours of dying? Are their health and social problems at an end and shall we then give up the profession of geriatrics and confine our scientific interest to gerontology? Of course not, but it is likely that there will be fewer things to worry about for old people than is the case today, and fewer of them will be coming to the doctor's office for help. Even so, aging will still be aging, and a strange process posing problems for every human being, and perhaps the approach of medical specialists should become less reductionist and more general. They may wish to view the whole person rather than concentrating on the singularities of individual diseases.

The word "holistic" was invented in the 1920s by General Jan Smuts to provide shorthand for the almost self-evident truth that any living organism, and perhaps any collection of organisms, is something more than the sum of its working parts. I wish holism could remain a respectable term for scientific usage, but, alas, it has fallen in bad company. Science itself is really a holistic enterprise, and no other word would serve quite as well to describe it. Years ago, the mathematician Poincaré wrote, "Science is built up with facts as a house is with stones, but a collection of facts is no more a science than a heap of stones is a house."

The word is becoming trendy, a buzzword, almost lost to science. What is called holistic thought these days strikes me as more like the transition from a mind like a steel trap to a mind like steel wool. What is sometimes called holistic medicine these days, if it is anything, strikes me as an effort to give science the heave-ho out of medicine, and to forget all about the working parts of the body and get along with any old wild guess about disease. We need another word, a word to distinguish a system from the components of a system, and I cannot think of one.

So, at present, we should continue to look at aging from the point of view of a biomedical scientist, which is the reductionist approach. In this way, we can construct hypotheses about the possible mechanisms of such problems as senile dementia and set about looking for selective enzyme deficiencies or scrapie-like slow viruses in the brain. Or we can theorize about the failure of cell-to-cell signals in the cellular immune system of aging animals and examine closely the vigor of lymphocytes at all stages of development; in the end, we may find out what goes wrong in an aging immune system. And perhaps we have within our grasp the information needed for explaining bone demineralization and resolving the problem, and we can track the neural pathways involved in incontinence and may learn to do something about that. Given some luck and a better knowledge of immunology and microbiology, we ought to be able to solve the problems of rheumatoid arthritis and osteoarthritis once and for all. We are learning some facts about nutrition and longevity that we never understood before. But we will still have people who will grow old before dying, and medicine will have to learn more about what growing old is like. The behavioral scientists, the psychiatrists, the psychologists, the sociologists, the anthropologists, and probably the economists, too, will all have a part in the work that needs doing, obtaining data, piecing information together, trying to make sense of the whole out of separate parts of the problem; but their individual efforts, although useful, may not be enough.

I suggest that we need a reading list for all young investigators and physicians to consult at the outset of planning for their ca-

reers in the scientific study of aging. Young people find it hard
to begin constructing hypotheses without having the ghost of an
idea what it means to be old. To get a glimpse of the matter,
you have to leave science behind for a while and consult literature, not "the" literature, as we call our compendiums of research, just plain old pure literature. I have a few sourcebooks
in mind. At the top of my list is Wallace Stegner, a novelist as
good as or better than anybody around, who in 1976 wrote *The
Spectator Bird*, about a literary man and his wife in their late
sixties and early seventies. To qualify for my list, you have to
be old enough to know what you are writing about, and Stegner
was the right age for his book; his novel should be required
reading for any young doctor planning on geriatrics. Indeed, *The
Spectator Bird* is good enough to help educate any young doctor
for any career.

Number two on my list is Malcolm Cowley and his book of
personal essays called *The View from Eighty*. This book is better
than any textbook on medicine, infinitely more informative than
most monographs and journals on geriatrics. Cowley wrote with
the authority of a man who had reached eighty "on the run" and
just getting a second wind. He was attached to Bruce Bliven,
the former editor of *The New Republic*, and quotes Bliven, who
writes, "We live by the rules of the elderly. If the toothbrush is
wet, you have cleaned your teeth. If the bedside radio is warm
in the morning, you left it on all night. If you are wearing one
brown and one black shoe, quite possibly there is a similar pair
in the closet." Bliven goes on, "I stagger when I walk, and small
boys follow me making bets on which way I'll go next. This
upsets me; children should not gamble." Malcolm Cowley writes
in good humor, and most of the people he admires who have
reached their eighties and nineties seem to share this gift, but
the humor is not always as light as it seems. An octogenarian
friend of his, a distinguished lawyer, said in a dinner speech,
"They tell you that you lose your mind when you grow older,
but what they don't tell you is that you won't miss it very much."

Stegner's central character in *The Spectator Bird* recounts
crankily that among his other junk mail was a questionnaire from

some research outfit that was sampling senior citizens and wanted to know intimate things about his self-esteem. He writes: "The self-esteem of the elderly declines in this society, which indicates in every possible way that it does not value the old in the slightest, finds them an expense and an embarrassment, laughs at their experience, evades their problems, isolates them in hospitals and sunshine cities, and generally ignores them, except when soliciting their votes or whipping off their handbags and social security checks." A few other old people have written seriously about their condition with insight and wisdom. Florida Scott Maxwell, a successful British actress, a scholar, and always a writer, wrote: "Age puzzles me. I thought it was a quiet time. My seventies were interesting and serene, but my eighties were passionate. I grow more intense with age. To my own surprise, I burst out in hot conviction. I have to calm down. I am too frail to indulge in moral fervor." Living alone in a London flat after the departure of her grandchildren for Australia and nearing her nineties, she wrote: "We who are old know that age is more than a disability. It is an intense and varied experience almost beyond our capacity at times but something to be carried high. If it is a long defeat, it is also a victory, meaningful for the initiates of time, if not for those who have come less far." She also wrote: "When a new disability arrives, I look about me to see if death has come, and I call quietly, 'Death, is that you? Are you there?' and so far the disability has answered, 'Don't be silly. It's me.' "

It is possible to say all sorts of good things about aging when you are talking about aging free of meddling diseases. It is an absolutely unique stage of human life—the only stage in which one has both the freedom and the world's blessing to look back and contemplate what has happened during one's lifetime instead of pressing forward to new high deeds. It is one of the three manifestations of human life responsible for passing along our culture from one generation to the next. The other two are, of course, the children who make the language and pass it along and the mothers who see to it that whatever love there is in a society moves into the next generation. The aged, if they are listened to, hand along experience and wisdom, and this transference, in the past, has always been a central fixture in the body of any culture. We do not use this resource well in today's society. We tend always to think of aging as a disability in itself—a sort of long illness without any taxonomic name, a disfigurement of both human form and spirit. Aging is natural, as we say, just as death is natural, but we pay our respects to the one no more cheerfully than to the other. If science could only figure out a way to avoid aging altogether, zipping us all straight from the tennis court to the deathbed at the age of, say, 120, we would probably all vote for that. But even if this could be accomplished by science, which is well beyond my imagining for any future time, society as a whole would take a loss.

In my view, human civilization could not exist without an aging generation for its tranquillity, and every individual would be deprived of an experience not to be missed in a well-run world. Aging is not universal in nature; it is not even common. Most creatures in the wild die off or are killed off at the first loss of physical or mental power, just as our own Olympic gymnasts lose their powers in their late teens and tennis stars begin to drop off in their twenties, and virtually all of our athletes become old for their professions long before late middle age. Aging, real aging—the continuation of living throughout a long period of senescence—is a human invention and perhaps a relatively recent one at that. Our remotest ancestors probably dealt with their aging relatives much after the fashion of more primitive cultures—that is, by one or another form of euthanasia. It took us a long time and a reasonably workable economy to recognize that healthy, intelligent, old human beings are an asset to the evolution of human culture. It was a good idea, and we should keep hold of it; but if the concept is to retain its earlier meaning, we will have to find better ways to make use of the older minds among us.

We need reminders that exceedingly useful pieces of work have been done in the past and are still being done by some extremely able people, sometimes in good health and sometimes in bad. Johann Sebastian Bach was relatively old for the eighteenth century when he died at age sixty-five, but he had just discovered a strange, new kind of music and was working to finish the incorporation of the art of the fugue into an astonishing piece based on the old rules but turning the form of composition into the purest of absolutely pure music. Montaigne was younger at his death at fifty-nine, but that was an old age for the 1500s. He was still revising his essays and making notes for a new addition for which he had chosen the appropriate title "He Picks Up Strength As He Goes." And in our own time, Santayana, Russell, Shaw, Yeats, Frost, and Forster—and I could lengthen the list by the score—were busy thinking and writing their way into their seventies and eighties and beyond. The great French poet Paul Claudel wrote on his birthday: "Eighty years old. No eyes left. No ears, no teeth, no legs, no wind, and, when all is said and done, how astonishingly well one does without them."

Of all the things that can go wrong in aging, the loss of the mind is far and away the worst and most feared. And here, I believe, is the greatest of all opportunities for medical science in the improvement of the human condition. I think that most aging people would willingly put up with all the other inconveniences of age, the awkwardness, the enfeeblements, even the assorted pains and aches, in trade for the assurance of hanging onto their minds. I cannot think of a higher priority for biomedical science today, and I believe that most younger people, now at no threat from Alzheimer's disease or any other kind of dementia, would agree with the priority. As matters stand today, with the increase in public awareness of this disease, whole families are beginning to worry about the problem, scrutinizing their parents and grandparents for any telltale sign of mental failure and wondering whether and when the entire family will topple in the devastating aftermath of this one disease. We need more and better research on the aging brain, on the biochemical and structural changes associated with dementia, on strokes and their prevention, on the slow viruses, and on autoimmune mechanisms. We need better ways of looking after afflicted patients than have yet been devised and more improvements in our facilities for institutional and home care and more help for the families. But, most of all, if we hope to rid ourselves of this disaster, we need research, good, old-fashioned, obsessive, reductionist science.

So I conclude on the same note of qualified optimism with which I began. If we keep at it, sticking to the facts at our reductionist best, we should be able to move gerontology onto a new plane among the biomedical sciences. The odds on success are very high indeed, provided that international cooperation in basic biomedical science is sustained and fostered in the years ahead. The odds on normal aging are already better than ever before in human history. With a lot of work and a lot of scientific luck, the odds can eventually become wholly on our side and medicine will have earned its keep.

Obligations

There are now approximately 4.5 billion members of our species alive, and sometime within the next half century that number will be almost surely double. Something around one-third of us, residing in modern, industrialized societies, enjoy what we should be calling reasonably good health, living out almost the estimated maximum life span for normal human beings. The rest, the majority of mankind, the citizens of impoverished nations, have less than half a chance at that kind of survival, dying earlier and living miserably for as long as they do live, threatened by constant hunger and an array of debilitating diseases unknown to the lucky third.

The central foreign policy question embedded in these loose statistics is the obvious one: what should the relatively healthy 1.5 billion human beings be doing to bring the other 3 billion into the twentieth (or twenty-first) century? I shall take it as given that there is an obligation of some sort here.

It is, in the first place, a moral obligation, but one driven by deep biological imperatives as well as by our conventional, cultural view of human morality. We are, like it or not, an intensely, compulsively social species. The reason we have survived thus far, physically fragile when compared with most other mammals, prone to nervous unsettlement by the very size and complexity of our forebrains, and competitively disadvantaged by our long periods of absolutely vulnerable childhood, is that we are genetically programmed for social living. I can no more imagine a solitary, lone human being, totally and permanently unattached to the rest of mankind, than I can envision a single termite making it through life on its own, or a hermit honey bee. What holds us together in interdependent communities is language, for which we are almost certainly as programmed by our genomes as songbirds are for birdsong.

I do not mean to suggest that we are very good at this, nor that we have been successful up to now. If that were the case, we would not have swarmed around the earth in our present dense masses, increasing our population by logarithmic increments up to today's risk of crashing, in the fashion of the last several centuries. We are fairly good at family life, allowing for the fact that some families drive their members, literally, crazy. Each one of us has a circle of close friends, trusted and even loved by them, and each of those friends has another circle, and you would think the expanding circles would extend in waves to include everyone, but it is not so. We have succeeded in working out long periods of survival in tribal units, allowing for the tendencies of tribes to make war against each other. It was in the invention of nation-states that we began to endanger our place in nature, by the implicit violation of all rules of social interliving. Instead of developing as a homogeneous unit of social animals, like an expanding termitarium, we took to splitting into colonies of ourselves and now all of them have become adversaries. Some, by luck or geography or perseverance, have turned out to be rich and powerful, others dirt-poor and weak, and here we all are, in trouble. Mankind is all of a piece, a single species, and our present situation will not do.

The only excuse I can make for us is that we are new at the game and haven't yet learned it. It is a mistake to think that the cultural evolution of humanity has been in any way analogous to biological evolution. We haven't been here long enough to talk about our living habits in the terms used by paleontologists and geologists. In the long stretch of epochs called time by the earth scientists, the emergence and development of social humanity could have begun only a few moments ago. We are almost brand-new. It may even be a presumption to say that we are already a juvenile species. The life of the earth is almost 4 thousand million years, and the evolution of species is recorded in spans of many million years each.

We may all be going through a kind of childhood in the evolution of our kind of animal. Having just arrived, down from the trees and admiring our thumbs, having only begun to master the one gift that distinguishes us from all other creatures, it should perhaps not be surprising that we fumble so much. We have not yet begun to grow up. What we call contemporary culture may turn out, years hence, to have been a very early stage of primitive thought on the way to human maturity. What seems to us to be the accident-proneness of statecraft, the lethal folly of nation-states, and the dismaying emptiness of the time ahead may be merely the equivalent of early juvenile delinquency or the accidie of adolescence. It could be, as some are suggesting, that we will be killed off at this stage, that what we are living through is endgame, but if so we will be doing it ourselves, and probably by way of nuclear warfare. If we can stay alive, my guess is that we will someday amaze ourselves by what we can become as a species. Looked at as larvae, even as juveniles, for all our folly, we are a splendid, promising form of life and I am on our side.

I would feel better about our prospects, and more confident for our future, if I thought that we were going to solve the immediate problem of inequity. It is one thing to say that some of us are smarter and more agile than others, more skilled in the management and enrichment of our local societies, therefore bound to be better off in living. It is quite another thing, however, to say that there is anything natural or stable about a world society in which two-thirds of the population, and all the children of that two-thirds, have no real chance at human living while those of us who are well off turn our heads away.

This is not the time, in today's kind of world, for me to be writing about equity in terms of the redistribution of the world's money, nor, surely, is this the place. Nor is this a matter that I possess qualifications for talking about, much less thinking about. But I do see the possibility, at least in technological terms, for doing something about the existing gross differences in the health of peoples in various parts of the earth. Moreover, I believe that this country, and the other countries like it in the so-called industrialized world, are under a moral obligation to do whatever they can to change these inequities, simply because we are members of a social species.

There are also, I suspect, obligations of a political nature, with substantial stakes of self-interest in having a stable, predictable world. The disease problems in the underdeveloped nations are, in part, the result of poverty and malnutrition, and these in turn are partly the result of overpopulation. But the problem turns itself around: the overpopulation is, in part, the result of the disease, poverty, and malnutrition. To get at the situation and improve it, something has to be done about all of these, in some logical sequence. It is not one problem, it is a system of problems. To change it without making matters worse will not be easy. Making well-thought-out changes in living systems is a dangerous business. Fixing one part, on one side, is likely to produce new and worse pathological events miles away on the other. The most dangerous of all courses is to begin doing things without even recognizing the existence of a system, and in this case a system in which all people, including the citizens of this rich country, are working parts.

If we should decide to leave matters roughly as they are, and to let nature simply take its course, it is hard for me to see how the future events can be politically acceptable, never mind morally. If a majority of human beings are to continue dying off before having a chance of life, succumbing to diseases that are, at least in principle, preventable, and in many cases dying of starvation with or without the associated diseases, this cannot be kept a secret. Television will be there, and as the disasters and the unhinging of whole countries of dying people become spectacular, the scrutiny by television will become closer and more continual. To say that this will have an unsettling effect on the viewing audiences in affluent countries is to understate the likely reaction. Meanwhile, the efforts will intensify among the billions of afflicted people to get out of wherever they are and to cross borders into any place where they can sense food and a hope of survival. Those left behind will continue the tropical deforestation already in progress, extinguishing immense ecosystems upon which other species are dependent and causing global climatic changes beyond predicting, jeopardizing the very life of the planet.

The greatest danger of all is in our own response. Having let nature take its course, we may someday decide that the problem has become insoluble, that the people knocking at our doors with hands outstretched have become enemies at our gates, who can only be coped with by the traditional method of killing them. This scenario is by no means unthinkable, not after this kind of century. We might even persuade ourselves that this is natural behavior, in aid of the whole species. Other kinds of animals, less equipped with brains and technologies than we are, take steps of their own to reduce their numbers by "crashing" from time to time. A "crash" is a jargon term used by ecologists for the catastrophic events which follow, inevitably, when any species has overreached itself, outnumbered the sources of food in its ecological niche and overgrown its allotted space on the planet. But other creatures, the "lower" animals as we like to call them, do not crash selectively; they crash all at once, all together.

Within another century it is likely that we will have swarmed everywhere, pole to pole, covering almost every livable acre of land space and water space. Some people are even talking seriously of *space* space, theorizing about the possibility of launching synthetic cities and countrysides enclosed in huge vehicles to sail the galaxy and perhaps colonize other celestial bodies.

What, on a necessarily limited scale, can we do? Specifically, what should we be planning for the improvement of the health of the masses of people who are now condemned by circumstances to lives that are, in the old phrase, nasty, brutish, and

short? Is it possible to do anything, without running the risk
of still another expansion of the human population? If people
everywhere could become as reasonably healthy as we are in
the United States, with birth rates and death rates approximately
the same as ours, would the world become intolerably overpopu-
lated? Considering the alternative—a massive population explosion already under way and beyond control, based in large part
on the reproductive drive among people now deprived of almost
everything except reproduction itself—I am not at all sure. It
seems to me worth a try, and I am unable to imagine any other
course of action.

When we in the Western world use the word "health," we
mean something considerably more than survival and the absence
of incapacitating diseases. To be healthy, we count it necessary
to be happy and rich as well. But for the present discussion, I
prefer to keep to the old-fashioned meaning.

Let us assume, in a flight of imagination, an economic state
of affairs in which it is financially possible for the richer countries
to export replicas of their entire technology of medical care to
the poor nations. This would involve, I suppose, prefabricated
versions of the Massachusetts General Hospital and Memorial
Sloan-Kettering Cancer Center, to be installed in every major
city in middle Africa, Asia, and South America, plus their professional staffs, plus duplicates of any top-drawer, accredited American medical school. And money enough to sustain these
enterprises for, at the least, a period of twenty-five years. I
believe the net effect of this munificence would be zero, or something less than zero. The affluent and influential members of
whatever establishment exists in the local bureaucracy would
doubtless enjoy the new institutions, and would save the air fares
now needed to fly them to hospitals in London or New York,
but the masses of people, especially those crowded into the
slums or still living out in rural areas, would be entirely unaf-
fected, or perhaps even adversely affected, because of the investment of all available funds on technologies totally
inappropriate to their health problems.

We are worlds apart. In our kind of society, today's enormously expensive health care system was put together in the decades following World War II primarily to cope with the medical concerns of people in their middle years and old age. The improvements in the general health of our populations, which began in the nineteenth century, had by this time reached such a high level that premature death had become less of a genuine day-to-day event and more of a nagging, and to some extent, a neurotic anxiety. Our attention is focused on such diseases as cancer, heart disease, and stroke, but we do not have to worry about dying early from the things that are killing most people every day in the Third World.

There is no question that our health has improved spectacularly in the past century, but there is a running argument over how this came to be. One thing seems certain: it did not happen because of medicine, or medical science, or the presence of doctors.

Much of the credit should go to the plumbers and engineers of the Western world. The contamination of drinking water by human feces was at one time the single greatest cause of human disease and death for us, and it remains so, along with starvation and malaria, for the Third World. Typhoid fever, cholera, and dysentery were the chief threats to survival in the early years of the nineteenth century in New York City, and when the plumbers and sanitary engineers had done their work in the construction of our cities these diseases began to vanish. Today, cholera is unheard of in this country, but it would surely reappear if we went back to the old-fashioned ways of finding water to drink.

But long before plumbing, something else happened to change our health prospects. Somehow, during the seventeenth and eighteenth centuries, we became richer people, here and in Europe, and were able to change the way we lived. The first and most important change was an improvement in agriculture and then in human nutrition, especially in the quantity and quality of food available to young children. As our standard of living improved, we built better shelters, with less crowding and more protection from the cold.

Medicine was only marginally involved in all this. Late in the nineteenth century the role of microbial infection in human disease was discovered, epidemiology became a useful science, chlorination of our water supplies was introduced, and quarantine methods were devised to limit the spread of contagions. The doctors had some voice in these improvements, but they did not invent the technologies. Medical care itself—the visits by doctors in the homes of the sick and the transport of patients to hospitals—had no more than a marginal effect on either the prevention or reversal of disease during all the nineteenth century and the first third of the twentieth. Indeed, during most of the centuries before this one, doctors often made things worse whenever they did anything to treat disease. They bled seriously ill patients within an inch of their lives and sometimes beyond. They administered leeches to draw off blood, spread blistering ointments over all affected parts of the body, purged the bowels with toxic doses of mercury, all in aid of eliminating what they called congestion of diseased organs, a figment of Galen's first-century-A.D. imagination.

In retrospect, there is nothing puzzling about the stunning success of homeopathy when it was introduced by Samuel Hahnemann in the mid-nineteenth century. Homeopathy was based on two notions, neither of them supported by any kind of science, both pure speculation on Hahnemann's part. The first was what he called the Law of Similars, that "like cures like." If a drug caused symptoms resembling those of a disease—fever or vomiting, for example—then that drug should be used for treating that disease. But it was his second notion that assured his success and that of his practice: the drugs should be given only in fantastically small amounts, diluted out to one part in 10 billion or more. In effect, homeopathic therapy was no therapy at all beyond reassurance, and a great many patients were thus protected against the conventional medicine of the day. No doubt they felt much better, and had a considerably improved prospect of recovery.

It was not until the early twentieth century that anything approaching rational therapy emerged for human disease, and it

was not until the middle of the century that we came into possession of rational and powerful technologies for the treatment and
prevention of infection on a large scale. Simultaneously with the
development of antibiotic therapy—and in large part because of
it—surgery underwent a comparable revolution. In the years
since, surgical techniques have become vastly more sophisticated
and powerful. Along with the control of infection, the surgeons
have learned enough about maintaining blood volume and electro-
lyte balance so that open-heart surgery, organ transplantation,
the repair of minute blood vessels, the replacement of severed
limbs, and extensive procedures for the removal of cancers once
considered unapproachable have become everyday, routine
procedures.

Now, can we package all this technology up, and send it along
to the impoverished nations? Should we? Would it be useful, or,
as the phrase has it these days, would it be cost-effective? I
think not.

I believe that what is needed for the health of Third World
societies is the same base of general hygiene that was put in
place in America and Europe before the introduction of modern
medicine. Unless this is done first, the adding on of our highly
expensive and sophisticated technologies simply cannot work.

The people who are dying prematurely in Central and South
America, and in most of Africa, and large parts of Asia, have a
different set of problems. They must raise their families in the
near certainty that half or more than half of their children will
die in infancy and early childhood, and accordingly they produce
as many as they can as early as they can. The losses are mostly
due to diarrheal diseases, caused by contaminated water supplies
and by faulty hygiene. The vulnerability of young children to
lethal infections is enhanced by inadequate food, and to some
extent as well by inadequate information about the selection of
food for young children. Aside from infant and child mortality
from infection and malnutrition, the major health problem for
people living in tropical and subtropical regions is parasitic
disease.

Here is a set of health problems for which this country can

really make itself useful, and may indeed be in the process of doing so. We can perhaps do a certain amount in helping to provide today's methods for the prevention and treatment of parasitic disease, but I am obliged to say quickly that these technologies are only marginally effective at their best, and they involve formidable logistical problems in getting them into the regions where they are needed. There are even more difficult obstacles—bureaucratic, cultural, and financial—in seeing to it that sick people actually receive them. We should be trying harder, even so, to do what we can to help.

We should not be trying to transport our high-cost, middle-class, middle-age and geriatric health care systems to our poor neighbors to the south, nor to their poor neighbors in Africa and Asia. They cannot cope with the high and still escalating costs of our kinds of high technology, nor is it what they really need at this stage. Our system, at its present stage, is designed to assure our citizens a chance at old age. What they hope for is a better chance at life itself. If we want to be useful, as we should, we ought to find ways to transfer another kind of governmental instrument which was, in our own laboratory, essential for our protection against contagion, malnutrition, and ignorance about health. This was the local health department as we had it during the late nineteenth through the first half of the twentieth century. By this time, our own local health departments have shrunk to vestigial organs, run out of things to do, and may be at the edge of extinction. But organizations like these, when they are working at top speed and have more missions than they could possibly cope with, are precise models for what an underdeveloped country needs. Not the typical, highly centralized bureaucratic ministry in the capital city, not even the partly decentralized but still too large organizations resembling our state health departments. What I am talking about is the small, old-fashioned board of health, as local and autonomous as possible, overseeing at first hand the health affairs of a country, a town, or a string of villages. This is an instrument that has really worked in our own past.

A few countries in Africa already have networks of organizations somewhat like these, but they are generally underfunded and understaffed by adequately trained personnel. More centers are needed, and the means would readily be found for providing the professionals to do the work and to train up a cadre of local professionals. This, I suggest, is where we might come in. We have not forgotten how to manage a local health department, and we have the people who understand the business.

They are the nurses. The profession of nursing was founded by energetic young women who learned how to handle most of the problems that bring patients to a doctor's office or a hospital clinic. Many of these people, around the turn of the century, left the hospital setting and established themselves in the poorest neighborhoods of our cities. They were called public health nurses, or visiting nurses. Their professional descendants are still being trained (better trained, indeed, than ever before), and are called nurse practitioners or physicians' assistants, including now some highly motivated young men as well as women.

If our government *does* intend to be helpful to its impoverished neighbors in the field of health, I do not believe that deploying American-trained physicians and surgeons in large numbers will meet the needs of foreign societies, even if it could be done. The real health problems are, in a sense, too fundamental for the products of our medical schools, most of them trained at very high cost for at least twelve years in universities and teaching hospitals before being certified, elaborately armed with medicine's highest technologies and aimed toward specialty practices. Very few of these people are either professionally or temperamentally prepared to cope with the day-to-day health problems of an impoverished and primitive society.

Nurses, by and large, come into their profession because of a straightforward ambition to be useful and helpful, and they hope to do this by simply "looking after" people who need them. With two or three years of nursing education (usually laid on these days after two years of undergraduate college education), they can take on the role once played by old-fashioned family doctors in our own rural communities, and they can do some things a good deal better. They can educate people who lack any understanding of hygiene and nutrition, they can organize systems for immunization of whole communities, and they can diagnose, or learn to diagnose, the endemic diseases of the region. When new drugs become available for the treatment or prevention of parasitic diseases, it is nurses who will be best equipped to see to it that they are properly employed in the field. They are perfectly capable of diagnosing and treating the common bacterial infections with the appropriate antibiotics.

The nurses in this country are also obliged, in the course of their training and early professional experience, to become good administrators. As a physician, I would have no qualms at all in seeing nurses take charge of the running of local health departments.

Hospitals are needed as well, of course, but not on anything like the scale in this country or Europe. A modest-sized network of small regional hospitals, designed after the fashion of Scotland's cottage hospitals, would be valuable if staffed by a limited complement of physicians and surgeons. Some of these professionals already exist in the countries concerned, and more could be trained in this country if we would turn our minds to it. The present difficulty is that they lack hospital opportunities and adequate incomes in their own countries, and tend to emigrate whenever the opportunity arises. An investment in hospital construction and maintenance is obviously necessary, but only on a small scale compared with what we do in this country.

These are what are needed: sanitation, decontaminated (or better still, uncontaminated) water supplies, antibiotics and vaccines and a distribution system assuring access to these things throughout the population, a chance at access to whatever new agents turn up for treating parasitic infections, plus a network of small hospitals with professional competence in primary medical care, plus a corps of visiting nurses, all of it run by nurse practitioners and physicians' assistants trained at the outset in this country and its affluent neighbors—to be later replaced or succeeded by similarly trained nurses from within the developing nations. These are the requirements for raising the standards of public health.

It must seem that what I am asking for is a new version of the Peace Corps, but one that must be considerably larger, drawing its professionals from all industrial nations rather than just the United States, and concentrating its attention on hygiene, infectious diseases, and nutrition. China has a health care system somewhat like this already in place.

The rewards for the professionals are there to see, and I need not dwell on them. It is not given to many young people to feel useful in society, and this is incentive enough for those chosen to serve overseas. But the nursing profession is in trouble here at home because of the derisory salaries paid, and to bring more bright young people into nursing schools from high schools and colleges will take more money than we have thus far been willing to pay. This is as true for Europe and the United Kingdom as here at home.

So one objection to what I hope to see happen is money, but it is not a vast sum, considering our other expenditures on our connections with the Third World, from defaulted loans to military apparatus. Money, in any case, is not the chief problem.

Our real contribution, for which we already possess the facilities and talent needed, will be in research. All of the diseases in question represent problems which are essentially unsolved, beyond knowing the taxonomic names of the parasites involved. Our methods for dealing with parasites are really quite primitive when compared with the technologies we have for treating and preventing bacterial and virus infections. Many of the chemicals commonly employed are nearly as toxic for the host as for the parasite, and the few effective ones—such as the current antimalarial drugs—are agents which the parasites quickly learn to resist. There is an enormous amount of pharmacologic and immunologic work still to be done.

The sheer numbers involved in the burden of parasitic disease seem overwhelming. Amebiasis affects 10 percent of the world's population, most of it in the Third World. The population at risk from malaria exceeds 1.2 billion, with an estimated 175 million people actively infected today. African trypanosomiasis (the curse of sleeping sickness) and American trypanosomiasis hold

70 million people at risk, and infect about 20 million right now. Schistosomiasis, worldwide, afflicts no fewer than 200 million people; filariasis and leishmaniasis 250 million; hookworm 800 million; onchocerciasis, a common cause of blindness in the tropics, 20 million.

One thing that should catch the eye immediately in this list of numbers is that these are not mortality figures, but morbidity statistics. The actual deaths that are caused each year by these diseases represent a very much smaller number, probably no more than 1 million deaths a year for malaria in all of Africa, for example. The disease burden is not so much a *dying* problem for the poor of the world as it is the problem of living on for a somewhat shortened life span with chronic, debilitating, often incapacitating disease. Trachoma does not kill patients, but blinds more than 20 million. There are about 15 million people with leprosy, one of the most chronic of all diseases. Malaria and schistosomiasis, which between them affect a billion and a half of the world's population, are vastly more important because of the human energy that is drained away each year than for the lives lost.

This provides an answer, of sorts, to the question I raised earlier: will solving the disease problem in the developing nations simply increase their populations to intolerable levels and make matters worse? It is probably not so. What might be accomplished would be the prospect of reducing the incidence of chronic invalidism, and vastly increasing the energy and productivity of billions of people.

Rough calculations can be made of the human energy costs entailed in certain diseases. A single day of malarial fever consumes more than 5,000 calories, for example. It has been estimated that this one disease represents a loss of about 20 percent of the total energy yield from grain production in the societies affected.

For many years, parasitic diseases have been thought of as problems unapproachable by real science, only to be dealt with by empirical and often exotic therapies. This view is changing rapidly. The cell biologists have recently learned how to cultivate

malarial parasites, the immunologists are fascinated by their surface antigens, and the molecular biologists are now about to
clone the genes responsible for the surface markers by which the
parasites protect themselves against the infected human host.
This means that a vaccine against malaria can now be thought of
as an entirely feasible prospect for the near future. The trypanosomes are becoming objects of fascination in contemporary genetics research because of their remarkable capacity to change
their surface antigens whenever the host begins to mobilize an
immune response, and the genes responsible for these evasions
are already being studied at first hand. If a vaccine to prevent
trypanosome infection in humans and farm animals can be devised, thus eliminating African sleeping sickness, this step alone
would open up for agriculture a fertile African area the size of
the United States which is now uninhabitable. My guess is that
parasitology will soon become one of the most active fields in
advanced biological science, and we should soon be finding ourselves in possession of an array of brand-new technologies for
both immunization and treatment.

Basic research on tropical infection and parasitism will be of
enormous benefit for the health, welfare, survival, and economic
productivity of the impoverished countries, but there is another
area of science which can be of equal importance in the long
term. We are just entering a new scientific frontier in agriculture,
thanks to the recent advances in molecular genetics and the
recombinant DNA technique. There is now a real possibility that
genetic manipulation can be used to transform the stress toler-
ance and disease resistance of current crops and grasslands. It
has been predicted by Frank Press, president of the National
Academy of Sciences, that "some 40 percent of the world's un-
cultivated but potentially productive land can be brought into
production," if fundamental problems in plant genetics can be
solved.

Here is also an opportunity for the Third World nations to
begin developing their own science base in biological science and
biotechnology. This matter has been the subject of a wrangling
debate within the United Nations Industrial Development Organi-

zation (UNIDO). At a meeting in Madrid in early September, 1983, ministerial delegates from twenty-five countries formalized an agreement creating, on paper, an international center for research and training in biotechnology, but they were unable to agree on a site for the center. India, Pakistan, Thailand, Tunisia, Bulgaria, Italy, Spain, and Belgium each proposed themselves as host countries. The discussion broke down in an argument over the question of the attractiveness of the center's location to world class scientists. There was the predictable polarization: the representatives of the Third World countries insisted that the center should be based somewhere in their region, while the American, British, French, West Germans, and Japanese spokesmen were uniformly negative. The name of the center was agreed upon, but little else: it will be called the International Center for Genetic Engineering and Biotechnology.

I should think that the dispute could be made less political and more scientific if the objectives of the proposed center were narrowed down and focused sharply on one high-priority area of research. There is little need in the Third World, at the moment anyway, for a major biotechnology research installation doing R and D on genetic engineering across the board. The manufacture for profit of products like growth hormone, interferon, insulin, or industrial enzymes is unlikely to be of much use to the economies of such societies. On the other hand, the application of genetic manipulation to agricultural research would be directly relevant. Moreover, the logical place to do this work would be in the regions of the planet where, for a variety of reasons, agriculture is technically not feasible or inadequate. The potential crops and feed animals to be improved exist in India and Africa, not in Belgium. Indeed, rather than having just one center in a single impoverished country, it would be more useful to set up a network of collaborating agricultural research centers in various countries of the Third World.

The problems in agricultural research have become of engrossing intellectual interest to many scientists throughout the industrialized world, and I have no doubt as to the feasibility of recruiting investigators to centers where the regional problems are both novel and urgent. Indeed, there is already an informal establishment of excellent Third World scientists trained in Europe and the United States who have expressed enthusiasm for the installation of biotechnology centers in their home countries, and who are confident that these institutions can, in time, become centers of excellence. It is a very different thing from the past (and failed) attempts to introduce heavy industrialization in hopes of transforming a poor country's economy. The scientific improvement of agriculture, and as a result, the transformation of a society's nutrition, have a greater potential for the improvement of human health than any other aspect of modern technology.

In conclusion, the objection usually cited against proposals of this sort, privately if not publicly, is that the survival assured to more children and the longer life span assured to more adults could abruptly increase the population beyond the resources of any conceivable food supply. I do not believe this. My guess is that populations given some confidence that living itself is possible, and that live children are possible, would be stabilized as never before in their histories. With luck, birth control could be accepted as a necessity for living (as it is now in China), and the present disastrous upswing in population might begin to level off. Without such a change in basic health standards, the curves will surely keep ascending, straight up until the final crash. It is worth the risk, I believe, and more than worth the relatively modest investment of money and talent from our side.

But my final argument, in my last ditch, is the simplest, most primitive, and perhaps least persuasive of all in terms of foreign policy. We *owe* it. We have an obligation to assure something more like fairness and equity in human health. We do not have a choice, unless we plan to give up being human. The idea that all men and women are brothers and sisters is not a transient cultural notion, not a slogan made up to make us feel warm and comfortable inside. It is a biological imperative.

III

Comprehending
my Cat
Jeoffry

I plan to step away from my own area of scientific competence
and attempt a few more or less random forays into another professional domain. Taking my life in my hands, then, I intend to
say some things about the mind.

Not the mind in general, mind you, only a peripheral, fuzzy
aspect of it. Namely, what I have titled the Awareness of Nature,
a topic that is intended to sound more ambiguous than ambitious
because I have a particular, minor ambiguity in mind. It comes
in three parts, as three connected questions: First, what do we
really think about nature these days, what do we collectively
mean by the word "nature" (or does it have a collective meaning
for us)? Second, what does nature think about itself, and incidentally about us, which of course raises the side issue of whether
any part of nature other than us does what we would call thinking? And, third, what lies at the center, for us and all the rest
of nature to have our attention focused on, or is there indeed a
center?

Let me say at the outset that it is not my intention to be very
serious about these matters, or to try to go very deeply. I shall
instead stay very close to the surface of things, mostly on thin
ice, playing light hunches all the way.

First hunch: I believe we've lost sight of, lost track of, lost touch with, and to some almost measurable degree lost respect for nature in recent years, certainly during the last years of this regrettable century. Somehow, we have done this during the very period of history when we humans have been learning more about the detailed workings of nature than in all our previous millennia. But the more we learn, the more we seem to distance ourselves from the rest of life, as though we were separate creatures, so different from other irrelevant occupants of the biosphere as to have arrived from another galaxy. We assert a duty to run the place, to dominate the planet, to govern its life, but at the same time we seem to ourselves to be less a part of it than ever before.

We leave it whenever we can, we crowd ourselves from open green countrysides on to the concrete surfaces of massive cities, as far removed from the earth as we can get, staring at it, when we must, from behind insulated glass, or by way of television half-hour clips. The urbanization of the world's human population is one of the strangest aspects of our species' recent behavior, and, along with our overpopulation, one of the most potentially disastrous. Lemmings go over cliffs, we move to town.

At the same time, we talk a great game of concern. We shout at each other in high virtue, now more than ever before, about the befoulment of our nest and about whom to blame. We have mechanized our lives so extensively that most of us live in the illusion that our only connection with nature is the nagging fear that it may one day turn on us and do us in. Polluting our farm-lands and streams, even the seas, worries us because of what it may be doing to the food and water supplies necessary for human beings. Raising the level of CO_2, methane, hydrofluoro-carbons, and whatnot in the atmosphere troubles our minds be-cause of the projected effects of climate upheavals on human habitats. These anxieties do not extend, really, to nature at large.

Nature itself, that vast incomprehensible meditative being, has come to mean for most of us nothing much more than odd walks in the nearby woods, or flowers in the rooftop garden, or the soap opera stories of the last giant panda or whooping crane. For some of us, the word "nature" now only reminds us of the northward approach, from Florida, of the Asiatic flying cockroach.

And yet, there is that great word, perhaps the most allusive and mysterious of all the words in English speech, with its roots in the Indo-European language of perhaps 35,000 years ago, and its direct progeny scattered through all the other languages derived from Indo-European. The original was *gen* or *gene*, meaning to give birth, to beget, to do all things associated with procreation and with the setting up of lineages. *Gen* moved into ancient Greek as *genos*, for race or family, and *gignesthai*, to be born; into Latin as *gnasci*, then *natus*, to be born. Along the way, cognate words were formed from the first root with the meaning of nature itself, the whole living world, and with all the things we truly believed were the essential parts and attributes of nature—natural, for example, human nature, good nature. And there, side by side, the other *gen* roots: gentle, generous, genetic, genital, genial, ingenious, benign. And, on another branch of the same tree, from the same root *gen*: kind, kin, kindly, even kindergarten. How could the long memory of such a word as nature, resonant with all such common meanings, be losing so much of its meaning?

Well, it must by now be obvious that I have not been talking about the people in this city or this country, or any country. Of course, I am talking about myself, my own misgivings about what is happening to my own feeling for nature. It is my own awareness of nature that troubles me today. I am talking to myself, from my own couch, in therapy. Waiting for an equivalent of transference, from one part of my mind to another. Or, more precisely, from one part of my brain to another.

That's better, I think. Moving the problem this way, from my mind to the real thing, my brain, opens new ground for my relief. How is it with all the other brains? What is going on in the minds of all the others, including our nonhuman relatives in the biosphere? What are they thinking, what have they been thinking? This last is, in most biological quarters, an outlandish question, even an impermissible one, to which the quick and easy answer is nothing, or almost nothing, or certainly nothing like thought as we use the word. None of them have real thoughts, none can foresee the future or regret the past, none are self-aware, except the super-primate humans, us.

And, by the way, how do I presume to mention them as our relatives? Am I talking just about our near cousins the chimpanzees and baboons, or about everyone? And if everyone, why call them relatives?

That one is easy. There is an Australian rock, pushing 3.7 billion years in age, containing the only relic of our original past, our indisputable Ur-ancestor, our n-granduncle, from which we and all the other parts of life in the biosphere are the lineal descendants. It was, beyond argument, a bacterial cell. For the next approximately 2.5 billion years, approximately two-thirds of the earth's total life span to date, the earth had no living occupants except the bacterial progeny of that original cell. The earth itself formed about 4.7 billion years ago, and it took a billion odd years for that firstborn. Then, soon, they were all over the place, in every conceivable niche, endlessly busy with their own version of evolution, getting things ready (including the invention of a livable, breathable atmosphere containing oxygen) for creatures like us. Their ways of evolution were as doctrinally Darwinian as ours, but perhaps quicker and more efficient and with many more options for diversity. Although they appeared to be diverging down the years into the numberless variants which we regard as separate species today, the term species may not have the same meaning in their kingdom. They succeeded in developing a set of mechanisms for interbreeding, passing bits of DNA across what we would be regarding as species barriers, engaged in such perpetual irrepressible interbreeding that Sonea and Panisset have seriously proposed that *all* bacteria, whatever their forms, are the actual equivalent of a tremendous clone, all one huge, loosely assembled organism.

Also, during all those years, the bacteria worked out elaborate schema for interliving and collaboration, even for an activity which I choose, for the moment, to take as a small but essential manifestation of thought. Not awareness, in any sense, God knows not with any process that anyone could call consciousness. The bacteria constitute by far the greatest living mass of tissue on and beneath the face of the earth, all interconnected and interdependent, communicating all day and all night in the most intimate exchange of information known to biology, the transfer of DNA from cell to cell, but this is not thought. It is an information system, but not what we can imagine as a mind.

But it is *part* of that process, partway along. Any time you have a system made up of huge numbers of communicating "agencies," continually engaged in a systematic mechanism for passing information back and forth and all around, you may have the makings of an intelligence. The bacterial kingdom is indeed such a system, and bacterial plasmids may be taken as metaphors for tiny ideas. Recently, when the notion of how to resist certain antibiotics popped into the plasmids of bacteria in some New York City hospitals, in almost no time the news had spread like gossip to similar bacteria in Peru, Australia, and Japan. Bacterial viruses can do this, carrying information from place to place, quicker than overnight mail. In a large-enough, coherent-enough system—and the world of bacteria is composed of trillions upon trillions of bits—the net result of this network might be as efficient as today's technology for what is called artificial intelligence.

Very early on, maybe 2 billion years ago, the bacteria began organizing themselves in diverse communities, rather like cities. The fossil relics are today's spectacular stromatolites, and the living equivalents are the algal mat found in tidal marshes everywhere. These are made up of multiple condominiums of microorganisms, stacked on each other like the floors of dense apartments with billions of tenants, each layer providing something indispensable in the way of metabolic products for the next story up, and making its living from other products seeping up from the floor below. These structures are genuine communities, living equably and congenially, in a world of economics not based on predator-prey relationships but a fair exchange. Instead of armaments for the taking and holding of territory, there are chemical signals continually emitted, some calling for more nutrients from the neighboring populations, some announcing the local geometry of borders to limit encroachment by neighbors, some serving as irresistible attractants.

Any living system dependent on this extensive degree of intimacy and interliving qualifies for the term "symbiosis," and most of the rules and regulations for symbiotic associations were laid out during the 2.5 billion years of evolution when there was nothing but microbial life on the planet. It is perhaps because of the direct inheritance of these rules, and the elaborate system of signals for cooperation, that the phenomenon of symbiosis evolved, as I believe it has, into the dominant mode of life in today's biosphere. Even including, with all appropriate reservations and misgivings, the society of us, the humans.

There now seems little doubt that the first nucleated cells came into being around 1 billion years ago as the result of symbiosis involving a pair, maybe a small committee, of previously free-living bacteria. One or more of these became the nucleus, another assumed the task of oxidative energy and turned into the ancestor of all mitochondria. Still another, a photosynthetic bacterium, perhaps the likes of today's blue-green algae (cyanobacteria), became the earliest chloroplast for the lineage of all green plants to come. Lynn Margulis has proposed that spirochetes joined some of the earliest committees as the antecedents of the cilia on our kinds of cells.

Scientists have identified a number of chemical molecules still in use for signaling among bacteria of various sorts which seem to have been inherited and put to other uses in multicellular organisms like us. A molecule remarkably like insulin in its properties is one of these; others are the small peptides now employed for regulating the growth and interaction of cells within our brains.

We owe a great deal, perhaps everything, to those countless generations of bacteria and their attendant viruses, and we should keep them in mind as we count our years. We have had a very brief run so far, only a million years or so as a vague taxonomic entity among the primates, probably only a few-score thousand years as the particular species identified by possession of genuine language. The bacteria, in contrast, have been at it for 3.7 thousand million years, real survivors, adaptors, cooperators, inventors. We owe them, for our benefit, much closer study and, as well, something like respect. Or, at the very least, much more thought. Their existence, and the plain fact that somehow or other they are our ancestors, strikes me as the strangest of all strange things in nature, and something to puzzle over forever.

We are probably the only species capable of acquiring this kind of information and then thinking about it. But it is surely not true that we are, as biological scientists and psychologists once asserted, the only species capable of thinking. It is my private view that there are many different ways of thinking, and an infinity of things to think about, and varieties of thought are going on all over the place. I mentioned awhile back that the kingdom of bacteria (with their viruses, plasmids, and other "small replicons") is engaged in something like artificial intelligence, due to the immense amount of information continually transmitted among living complex entities of the type that might fairly be compared to what Marvin Minsky has termed "agents" in an idealized AI machine. Move up on the evolutionary ladder, into the world of early invertebrates, and there are small, apparently mindless organisms like the sea slugs in Eric Kandel's laboratory, fully capable of memory, long-term as well as short-term. Not yet real thinking in our terms, but surely the most fundamental and necessary component of thought; I cannot imagine any sort of thinking in the absence of some sort of memory. And then, move up from there to my cat Jeoffry. As you come "higher," as one says, in the direction of us, past the more or less established familiar down the block all the way up to us, the real newcomers, science, technology, money and all, you do indeed run into what seems to me, inescapable: awareness along the way.

Trouble is, most of us, up here at the peak of evolution, cannot manage the awareness of our own awareness. We do a lot of thinking all day and (I am told) all night, but it is only on rare occasions that we run across a real new thought. It may even turn out, to our embarrassment, that some of our Asiatic neighbors have been right about this problem all along: that the awareness of awareness is only achieved when the mind succeeds in emptying itself of all other information and switches off all messages, interior and exterior, a state of mind for which the Chinese Taoists long ago used a term meaning, literally, "no-knowledge." With no-knowledge, it is said, you get a different look at the world, an illumination.

Falling short of this, as I do, and dispossessed of anything I could call illumination, it has become my lesser satisfaction to learn secondhand whatever I can, and then think about, firsthand, the behavior of other kinds of animals.

Crickets, for instance, and the thought of their unique, very small thoughts, principally about mating, and about bats, but also, perhaps, about the state of cricket society. The cricket seems to me an eminently suitable animal for sorting out some of the emotional issues bound to arise in any consideration of animal awareness. Nobody, so far as I know, not even a nineteenth-century minor poet, could imagine any connection between events in the mind of a cricket and the mind of a human. If there was ever a creature in nature meriting the dismissive description of a living machine, mindless and thoughtless, the cricket qualifies. So, in talking about what crickets are up to when they communicate with each other, as they unmistakably do, by species-unique runs and rhythms of chirps, there can be no question of anthropomorphism, that most awful of all terms for the deepest error a modern biologist can fall into.

If you reduce the temperature of a male cricket, the rate of his emission of chirping signals is correspondingly reduced. Indeed, some of the earlier naturalists used the technical term "thermometer crickets" because of the observation that you can make a close guess at the air temperature in a field by counting the rate of chirps of familiar crickets.

This is curious, but there is a much more curious thing going on when the weather changes. The female crickets in the same field, genetically coded to respond specifically to the chirp rhythm of their species, adjust their recognition mechanism to the same temperature change and the same new, slower rate of chirps. That is, as Doherty and Hoy wrote on observing the phenomenon, "warm females responded best to the songs of warm males, and cold females responded best to the songs of cold males." The same phenomenon, known as "temperature coupling," has been encountered in the mating behavior of grasshoppers and tree frogs, and also in the flash communication systems of fireflies. The receiving mind of the female cricket, if you are willing to call it that, adjusts itself immediately to match the sending mind of the male. This has always struck me as one of the neatest examples of animals adjusting to a change in their environment.

But I started thinking about crickets with something quite different in mind, namely bats. It has long been known that bats depend voraciously on the nocturnal flights of crickets and moths, which they detect on the wing by their fantastically accurate ultrasound sonar mechanism. What should have been guessed at, considering the ingenuity of nature, but recognized only recently, is that certain cricket species, and certain moths, have ears that can detect the ultrasound emissions of a bat, and can analyze the distance and direction from which the ultrasound is coming. These insects employ two separate and quite distinct defensive maneuvers for evading the bat.

The first is simply evasion—flying to one side or the other, or merely dropping to the ground. This is useful behavior when the bat signal is coming from a safe distance, twenty to thirty meters away. At this range, the insect can detect the bat, but the bat is too far off to receive the bounced ultrasound back to his own ears. So the cricket or moth needs to do nothing more, at least for the moment, than swing out of earshot.

But when the bat is nearby, three meters or less, the insect is in immediate and mortal danger, for now the bat's sonar provides an accurate localization. It is too late for swerving or veering; the bat can easily track such simple evasions. What to do? The answer has been provided by K. D. Roeder, who designed a marvelous laboratory model for field studies, including instruments to imitate the intensity and direction of bat signals.

The answer, for a cricket or moth or lacewing who hears a bat homing in close by, is chaos: instead of swerving or dropping, the insect launches into wild, totally erratic, random flight patterns, as unpredictable as possible. This kind of response tends to confuse the bat, and results in escape for the insect frequently enough to have been selected by evolution as the final, stereotyped, "last-chance" response to the threat. It has the look of a very smart move, whether thought out or not.

So chaos is part of the useful, everyday mental equipment of a cricket or a moth, and that, I submit, is something new to think about. I don't wish to push the matter beyond its possible significance, but it seems to me to justify a modest nudge. The long debate over the problem of animal awareness is not touched by the observation, but it does bring up the opposite side of that argument, the opposite of anthropomorphization. It is this: leaving aside the deep question as to whether the lower animals have anything going on in their minds that we might accept as conscious thought, are there important events occurring in our human minds that are matched by habits of the animal mind? And, surely, chaos is a capacious area of common ground. I am convinced that my own mind spends much of its waking hours, not to mention its sleeping time, in a state of chaos directly analogous to that cricket hearing the sound of that nearby bat. But there is a big difference. My chaos is not induced by a bat, it is not suddenly switched on in order to facilitate escape, it is not an evasive tactic set off by any new danger. It is, I think, the normal state of affairs, and not just for my brain in particular but for human brains in general. The chaos that is my natural state of being is rather like the concept of chaos that has emerged in higher mathematical circles in recent years.

As I understand it, and I am quick to say that I understand it only quite superficially, chaos occurs when any complex, dynamic system is perturbed by a small uncertainty in one or another of its subunits. The inevitable result is an amplification of the disturbance and then the spread of unpredictable, random behavior throughout the whole system. It is the total unpredictability and randomness that makes the word "chaos" applicable as a technical term, but it is not true that the behavior of the system becomes disorderly. Indeed, as Crutchfield and his associates have recently written, "There is order in chaos: underlying chaotic behavior there are elegant geometric forms that create randomness in the same way as a card dealer shuffles a deck of cards or a blender mixes cake batter." The random behavior of a turbulent stream of water, or of the weather, or of Brownian movement, or of the central nervous system of a cricket in flight from a bat are all determined by the same mathematical rules. Professor Jay Forrester, at MIT, encountered behavior of this sort in his computer models of large cities: when he made a small change in one small part of the city model, the amplification of the change, even a minor one, resulted in enormous upheavals in the municipal behavior at remote sites in his models, none of them predictable.

A moth or a cricket has a small enough nervous system to *seem* predictable and orderly most of the time. There are not all that many neurons, and the circuitry contains what seem to be mostly simple reflex pathways. In a normal day, one thing—the sound of a bat at a safe distance, say—leads to another, predictable thing: a swerving off to one side in flight. It is only when something immensely new and important happens—the bat sound three meters away—that the system is thrown into chaos.

I suggest that the difference with us is that chaos is the norm. Predictable, small-scale, orderly, cause-and-effect sequences are hard to come by, and don't last long when they do turn up. Something else almost always turns up at the same time, and then another sequential thought intervenes alongside, and there come turbulence and chaos again. When we are lucky, and the system is operating at its random best, something astonishing may suddenly turn up, beyond predicting or even imagining, and events like these we recognize as good ideas.

I am not sure where to classify the mind of my cat Jeoffry. He is a small Abyssinian cat, a creature of elegance, grace, and poise, a piece of moving sculpture, and a total mystery. We named him Jeoffry after the eighteenth-century cat celebrated by the unpredictable poet Christopher Smart in his "Jubilate Agno,"

in which the following lines appear, selected here more or less
at random:

For I will consider my Cat Jeoffry . . .
For he counteracts the powers of darkness by his electrical skin and
  glaring eyes.
For he counteracts the Devil, who is death, by brisking about the life.
For in his morning orisons he loves the sun and the sun loves him.
For he is of the tribe of Tiger . . .
For he purrs in thankfulness, when God tells him he's a good Cat.
For he is an instrument for the children to learn benevolence upon . . .
For he is a mixture of gravity and waggery . . .
For there is nothing sweeter than his peace when at rest.
For there is nothing brisker than his life when in motion . . .
For he can swim for life.
For he can creep.

I have not the slightest notion what goes on in the mind of
my cat Jeoffry, beyond the conviction that it is a genuine mind,
with genuine thoughts and a strong tendency to chaos, but in all
other respects a mind totally unlike mine. I have a hunch, based
on long moments of observing him stretched on the rug in sunlight, that his mind has more periods of geometric order, and a
better facility for switching itself almost, but not quite, entirely
off, and accordingly an easier access to pure pleasure. Just as
he is able to hear sounds that I cannot hear, and smell important
things of which I am unaware, and suddenly leap like a crazed
gymnast from chair to chair, upstairs and downstairs through the
house, flawless in every movement and searching for something
he never finds, he has periods of long meditation on matters that
I know nothing about.

His brain is vastly larger and more commodious than that of
a cricket, but I wonder if it is qualitatively all that different. The
cricket lives with his two great ideas in mind, mating and bats,
and his world is a world of particular, specified sounds. He is a
tiny machine, I suppose, depending on what you mean by machine, but it is his occasional moments of randomness and unpredictability that entitle him to be called aware. In order to achieve
that feat of wild chaotic flight, and thus escape, he has to make

use, literally, of his brain. The neurons and interconnecting fibers that govern most of cricket behavior are relatively simple, equivalent to direct sensory-motor reflex arcs, but this is not the case for his escape flight. For this, he must make use of a more complex system involving an auditory interneuron known as Int-1. When Int-1 is activated by the sound of a bat closing in, the message is transmitted by an axon connected straight to the insect's brain, and it is here, and only here, that the chaotic flight behavior is generated. This I consider to be a thought, a very small thought but still a thought. Without knowing what to count, I figure my cat Jeoffry, with his kind of brain, has a trillion thoughts of about the same size in any waking moment. As for me, and my sort of mind, I can't think where to begin.

We like to think of our minds as containing trains of thought, or streams of consciousness, as though they were orderly arrangements of linear events, one notion leading in a cause-and-effect way to the next notion. Logic is the way to go; we set a high price on logic, unlike E. M. Forster's elderly lady in *Aspects of the Novel* who was overheard saying, "Logic? Rubbish! How do I know what I think until I see what I say?"

I can acknowledge openly that my own mind is, at most times, a muddled jumble of notions, most of them in the form of questions, never lined up in any proper order to be selected and dealt with when time allows, most of the time popping into my head unpredictably and jostling against any other ideas that happen to be floating along, each new disturbance amplifying the disorder of all the others, creating new geometric shapes of chaos imposed on chaos.

I think I might be able to straighten things out, and restore something like order, or at least less noise, if only a few of the strangest puzzles that now keep me confounded could be settled, once and for all.

I can make a short list of some of these, but only with a cautionary note that this is today's list, and tomorrow it will undoubtedly be a different jumble.

High on my list is the most nagging of all questions: what on earth is the mind? Never mind the neat explanations involving all those columnar modules of cells in my cortex, all connected by intricate tracts of fibers with numberless others, wired with redundancy and set up with chemical modulators that have the effect of making certain subsystems more likely to be selected for memories, and all that. That line of study is fine in accounting for the general outlines of a structural design for assembling the billions of cells involved in a single, coherent message within the brain, and then connecting that message with other messages already embedded in the circuitry. I admire this line of study and am all for it, but no matter where it takes us in comprehending the hierarchical machinery of the brain, it still leaves the same infuriating question: even so, what is the mind? Is it something separate, some Cartesian disembodied entity that emerges at the top of all the machinery and then turns back to play it like an instrument, saying "Move that finger," or "Now think that thought"?

Or is consciousness something else, some sort of entity that comes along in sections of itself, depending on how many units of cells are at work, and in what part of the brain? Is a Princeton honey bee conscious of what is going on in its brain when it comes into the dark hive with the news that Professor James Gould has just begun to move the dish of sugar another 200 yards to the south-southeast, and if we fly out now to that site we'll be there before Gould gets there? In a termite nest, twelve feet tall and still growing, even agreeing as we must that there isn't enough brain in any single termite to achieve anything remotely resembling a thought, beyond maybe knowing that the local pheromone concentration means that it is time to pick up another fecal pellet and climb up to where the next arch in that part of the nest is ready for perfect turning, what can we say about the hundreds of thousands of termites all fitted together in the dark, all connected by touch and by smell, doing architecture and wood sculpture, and air-conditioning and military maneuvers and perpetual breeding in a beautiful kind of harmony? Is that collection of insect brains, each with a life span of only a few weeks within a live structure sixty years old, all linked together, a mind of some kind? Is an anthill or a beehive or a termitarium, as Marais and Wheeler and others have proposed, a "superorganism"? And if this is so, what of us? When we humans are collected together at close quarters all over the face of the earth, in angry mobs or in orderly but folly-prone nation-states or in auditoriums listening carefully and silently to the Late Quartets, are we engaged in a kind of collective thought that is a qualitatively different process from what we think is happening within our individual heads?

How does it happen that our bodies get all the necessary things done for us, or most of them anyway, without our conscious minds being required to intervene or even supervise the process? How would you feel if you were suddenly told, "There is your liver, right upper quadrant, now go ahead and operate it yourself"? Or your pineal gland, heaven help you? Or, worst of all possibilities, start running your own brain! But then, it has been reported that warts can be made to fall off under hypnosis, and blisters raised, and how does that work?

It is the easiest of things to lose control of your brain, or your mind, or whatever it is. If you give it its head, there is no telling where it will go next or what it will try to do. It is the puzzles that do it, and so far there is no way to fend the puzzles off. They come in from all sides. Personally, I use a method of my own for treating this condition, somewhat like homeopathy: introduce an artificial but still insoluble puzzle of your own, which will have the effect of silencing, or at least quieting down, all the noise. The tail of the dog.

My contrivance is the Fibonacci series of numbers. Once you allow this into your mind, always provided that you are not an educated mathematician, preferably like me not even qualified as an amateur, the effect is magical—for only a short time, of course, but these days even a brief run without chaos is a summer in the Swiss Alps.

The numbers are arranged in a series such that each member is the sum of the two preceding numbers. Thus it goes: 1, 1, 2, 3, 5, 8, 13, 21, 34, 55, 89, 144, 233, 377, 610, 987, 1597, 2584, 4181, 6765, and so forth into, presumably, infinity. Among the things to think about are that the ratio of any number to the preceding one is the famous irrational number 1.618034, while its ratio to the next one up is 0.618034. This relationship defines the so-called golden section, on which architecture has relied for its traditional aesthetic, and which defines the growth of trees, the replication of rabbits, the spirals of seeds in sunflowers and the fronds of pine cones, and even, as I have recently read in a learned paper by an incomprehensible (to me) Hungarian musicologist, the composition of Bartók.

Generations of amateurs have puzzled over the Fibonacci series and its endless ramifications, ever since Leonardo Fibonacci set it up in the thirteenth century. I have filled a notebook with my own "discoveries," all doubtless discovered by others a hundred or more years ago. Like this: 322 (or 321.9) multiplied by any number in the series will give (roughly) the twelfth number ahead in the series; the square root of 322 gives the sixth number ahead; the cube root gives three numbers ahead; the fourth root gives a number 3.3301 times the first number back, and so forth.

Add up all the digits in a Fibonacci number, and reduce this sum to a single digit, as follows: 10,946 (which happens to be the twenty-first number) = 20 = 2. Now do the same with the twelfth number back from 10,946, which is 34, or the twelfth ahead, which is 3,524,578, and you will find that these digits sum to 7. Thus, the digits in every pair of Fibonacci numbers, twelve apart, have the sum of 9, whether $7+2$, or $5+4$, or $6+3$, or $8+1$. Moreover, these reduced digits, when lined up, display a regular, absolutely reproducible cycle, every twenty-four Fibonacci numbers, no matter how big the numbers become.

Someone knows for sure why these things are so, and proper mathematicians can no doubt prove the empirical facts by equations, but I don't care. For my purposes, the Fibonacci series keeps my mind quiet—not at peace, for that would be too much to ask for, but quiet at least. And, I might add, something distantly related, as my mind sees it, to the sound of music.

Some people are critical of the drift of science these days, fearful of the future, apprehensive that we may have already learned too much about the interior workings of nature, worried that there may be some things just ahead that human beings should simply not be finding out, pieces of knowledge that we will be unable to handle at this stage in the development of our species. There are, it is asserted, certain things we are better off not knowing about.

Science, it is being said, has become an endeavor touched by *hubris*, that hybrid word from roots literally meaning outrage, later adapted from the Greek *hubris* to Latin *hybrida* for the meaning of a disastrous cross-breeding, now back to the original sense of *hubris* to signify so much of outrageous pride as to risk offense to the gods, or in other terms an offense to nature.

It seems to me that our problem is precisely the opposite: we are learning from science how little we know, how still less we understand, and how much there is to learn. If we are endangered by science it is because of the feckless technologies we have picked up as spin-offs from science, and when we develop such technologies—nuclear weapons as just one example among, I begin to fear, many others—we will blunder our way into trying to use them because of our ignorance about nature.

I will begin to feel better about us, and about our future, when we finally start learning about some of the things that are still mystifications. Start with the events in the mind of a cricket, I'd say, and then go on from there. Comprehend my cat Jeoffry and we'll be on our way. Nowhere near home, but off and dancing, getting within a few millennia of understanding why the music of Bach is what it is, ready at last for open outer space. Give us time, I'd say, the kind of endless time we mean when we talk about the real world.

Science and the
Health of
the Earth

A human embryo—or for that matter any kind of embryo—begins life as a single cell, which promptly starts dividing into progeny cells until a critical mass of apparently identical cells is formed. Then, as though a bell had sounded, it begins the process of differentiation, and cells with specialized functions for the future organism make their appearance and migrate to one region or another for the formation of tissues and organs. The current view is that the sorting out of cellular elements is organized and governed by a system of chemical signals exchanged among the cells. The nature of these signals and their specific receptors at each cell surface are presumably determined and controlled by the particular genetic instructions processed by that very first cell and passed along to its offspring.

It is not at all understood how this system works, even though we have a clear and detailed picture of the structural changes that take place at every stage of the process, and small bits of information about some of the chemical messages in the differentiation of certain cells. The phenomenon of embryologic development and differentiation is generally regarded as one of the two most profound unsolved problems in human biology—the other being the operation of the brain. In both cases, the core of the mystery is the cooperative and collaborative behavior of the cells themselves. The embryo develops from a single cell into an elaborately complex structure, a baby, made up of trillions of cells, each one specialized for doing what it is supposed to do and confining its activities to its designated anatomical area but kept in communication with all the rest by chemical advertisements. The brain is made up of billions of neurons arranged in wiring networks of a complexity beyond comprehension, but under the governance as well of chemical signals which regulate the firing and response of every cell.

The life of the earth resembles that of an embryo, and the life of our species within the life of the earth resembles that of a central nervous system. The earth itself is an organism, still developing and differentiating.

The planet formed as a solid oblate sphere and swung into its orbit roughly 4.7 billion years ago. Less than a billion years later, the first life appeared. We do not have an attested date for this, but we do have fossil evidence for the existence of chains of bacteria resembling streptococci some 3.7 billion years old.

Nobody knows how the first living thing was put together, although hypotheses abound. It is, however, a near certainty that it was a cell, and most likely a cell resembling one of today's bacteria. It might have been a virus, but if so it would have to have been a virus carrying genetic specifications for making a cell. The earth was still very hot in many places 4 billion years ago, and it is possible that life may have turned up first in a very hot place. Indeed, it makes it easier to account for so deeply improbable an event if we assume a high temperature. Most of our guesses about the origin of life postulate, necessarily, a series of random accidents—the presence of amino acids and precursors of nucleotides in the water covering most of the planet, the assembling of these building blocks into more complex nucleic acids and protein molecules under the influence of lightning or intense ultraviolet light, the formation of biological membranes to enclose the reactants, and presto, life. One trouble with the scenario is the time required for the right sequence of events to occur at what has always been assumed to be today's optimal temperature for life. Each step by itself would be almost inconceivably improbable, but if the events were to have occurred at very high temperatures, with everything greatly speeded up, a billion years doesn't seem so short a time. It now becomes a possibility, for speculation anyway, that the original ancestors of all the planet's life were bacteria of this sort.

We have inherited many, perhaps all, of our systems for intercellular communication from our bacterial ancestors. Investigators at the National Institutes of Health have recently discovered that certain bacteria manufacture protein molecules indistinguishable from insulin in their properties. Other microorganisms elaborate peptide messengers identical to those used by specialized cells in our bodies for the regulation of brain function and for switching on the activity of our thyroid, adrenal, ovarian, and digestive cells. We did not invent our steroid hormones; molecules like these were probably being made for other reasons 2 billion years ago by our bacterial forebears. Biochemically speaking, there is nothing new under the sun.

The level of oxygen in the atmosphere has gradually increased because of the steady increase in the earth's population of photosynthetic organisms—some of them still in their original bacterial form, the blue-green algae living in the waters of the earth, others now in the more complex and various forms of higher plants. From a level close to zero 3.5 billion years ago, oxygen now stands at 20 percent of the earth's atmosphere, all of it the product of life itself. There is another feature of that level of oxygen that seems to me equally remarkable: it stabilized at its present level around 400 million years ago, and it seems to be fixed there. It is a lucky thing for us, and for the life of the earth, that it did stabilize at that concentration. If it were to increase by more than two percentage points, most of the planet would ignite. If it were to decrease a few points, most of the life would suffocate. It is nicely balanced, against all hazards, at an absolutely optimal level.

Other gases in the atmosphere, including carbon dioxide, nitrogen, and methane, also seem to have been regulated and stabilized over long periods of time at optimal concentrations, despite the constant intervention of natural forces tending to push them up or down. At the moment, which is to say over the past century, the level of carbon dioxide has been rising slowly due to the heavy increase in the burning of fossil fuels, but the climatic consequences of this rise have not yet been observed.

Methane exists as a quantitatively minor constituent of the atmosphere, although it plays a crucial role in the interests of living things. Most of what is there is the product of life itself, the tremendous populations of methanogenic bacteria in the soil and water, in the intestinal tracts of ruminant vertebrates, and (a substantial source) in the hindgut of termites. How it is regulated so that the methane concentrations are everywhere fixed and stable is not known, but it is known that if the level were to decrease appreciably the concentration of oxygen in the atmosphere would begin to rise to hazardous levels. There is probably a feedback loop, in which methane serves as a regulator of oxygen and vice versa.

The mean temperature at the surface of the earth has also remained remarkably stable over stretches of geological time, perhaps due in large part to the relatively stable concentrations of carbon dioxide in the atmosphere. From time to time fluctuations have occurred, with the cyclical development of ice ages as the result, but over all time, the temperature stays much the same. This also suggests a regulatory mechanism of some sort, since the radiant heat coming from the sun has increased by approximately 30 percent since life began.

Lovelock and Margulis proposed in 1972 that life on the planet has been chiefly responsible for the regulation of that life's own environment. They postulated that the stability of the constituents of the earth's surface, and the pH and salinity of the oceans, are held more or less constant, and at optimal levels for life, by intricate loops of feedback reactions involving microbial, plant, and animal life. The concept is analogous to the phenomenon of homeostasis within a multicellular organism, as outlined by Claude Bernard and later elaborated by Walter Cannon to explain the stability of the internal environment of the human body. If one tissue component begins to change, other sets of components will respond by changing things back again to where they were.

The "Gaia Hypothesis," as Lovelock and Margulis termed their theory, plainly implies that the conjoined life of the earth behaves like a huge, coherent, self-regulating organism. It is a notion that has aroused both skepticism and antipathy within the biological community, especially among evolutionary biologists. They do not much like the name, for one thing, with its undertones of deity and deification. For another, they doubt its compatibility with the solid body of evolutionary theory. How could such a creature have evolved, they ask, in the absence of anything to be selected against? Moreover, they object to the idea that evolution can plan ahead for future contingencies—for example, the adjustment of atmospheric gases to provide optimal conditions for forms of life that have not yet come into existence. I will not deal with these objections here, except to remark that similar puzzles confront biologists who wish to explain the development of an embryo. I can find it in my imagination to suppose that once that first primordial cell came on the scene, equipped with the molecule of DNA for its replication but also, more importantly, for its progressive mutation into new cellular forms with new strategies for living, a *system* had come into existence. When a living system becomes sufficiently complex, it automatically provides a series of choices among strategies for future contingencies. When these turn up, it has the look of planning and purposiveness, but these are the wrong words. What we do not know enough about are the strategies never used, for contingencies that never turned up.

This is not to suggest that the earth's environment is perpetually benign, nor that accidents cannot happen. To the contrary, immense accidents are the rule, but they have been spaced far enough apart in geologic time so that whatever the damage, life itself has always had plenty of time to recover and reappear in a new abundance of more complex forms. The most devastating accident on the paleontological record was the mass extinction of the late Permian, 225 million years ago, when at least 50 percent of the marine fauna were lost. The catastrophe was the result of the coalescing of the world's continents into a supercontinent (now referred to geologically as Pangea), which eliminated most of the shallow sea habitats available to the marine creatures then alive. The second greatest extinction occurred around 65 million years ago, when half of the marine creatures and many terrestrial animals, including all the dinosaurs, simply vanished. Various explanations for this extinction have been proposed, including volcanic eruptions or an asteroid collision resulting in earthwide clouds of dust blotting out the sun for a long enough time to bring most plant and animal life to an end.

We are not finished with great extinctions. The current anxiety in some biological quarters is that the next one may be just ahead, and will be the handiwork of man.

At a national meeting of biologists and biogeographers held in Arizona in August 1983, the history and dynamics of extinction were the topics of discussion. The consensus was that the number and diversity of living species may be on the verge of plummeting to a level of extinction matching the catastrophe that took place 65 million years ago, and that this event will probably occur within the next hundred years and almost certainly before two hundred years. It will be caused, when it occurs, by the worldwide race for agricultural development, principally in the poorer countries, and by the appalling rate of deforestation. Although tropical forests cover only around 6 percent of the earth's land, they harbor at least 66 percent of the world's biota, animals, plants, birds, and insects. They are currently being destroyed at the rate of about 100,000 square kilometers per year. Elsewhere on the planet, urban development, chemical pollution (especially of waterways and shoreline ecosystems), and the steady increase in atmospheric carbon dioxide are posing new problems for a multitude of species. The animal species chiefly at risk for the near term is humankind. If there is to be a mass extinction just ahead, we will be the most conspicuous victims. Despite our vast numbers, we should now be classifying ourselves as an immediately endangered species, on grounds of our total dependence on other vulnerable species for our food, and our simultaneous dependence, as a social species, on each other.

But do not worry about the life of the earth itself. No extinction, no matter how huge the territory involved or how violent the damage, can possibly bring the earth's life to an end. Even if we were to superimpose on the more or less natural events now calculated to be heading toward a mass extinction the added violence and radioactivity of a full-scale, general nuclear war, we could never kill off everything. We might reduce the numbers of species of multicellular animals and higher plants to a mere handful, but the bacteria and their resident viruses would still be there, perhaps in greater abundance than ever because of the expanding ecosystems created for them by so much death. The planet would be back where things stood a billion years ago, with no way of predicting the future course of evolution beyond the high probability that, given the random nature of evolution, nothing quite like us would ever turn up again.

If the ecologists are right in their predictions, we are confronted by something new for humanity, a set of puzzles requiring close attention by everyone. It is something more than an international problem to be dealt with by the specialists in each nation who deal with matters of foreign policy. Human beings simply cannot go on as they are now going, exhausting the earth's resources, altering the composition of the earth's atmosphere, depleting the numbers and varieties of other species upon whose survival we, in the end, depend. It is not simply wrong, it is a piece of stupidity on the grandest scale for us to assume that we can simply take over the earth as though it were part farm, part park, part zoo, and domesticate it, and still survive as a species.

Up until quite recently we firmly believed that we could do just this, and we regarded the prospect as man's natural destiny. We thought, mistakenly, that that was how nature worked. The strongest species would take over. The weak would be destroyed and eaten or used in other ways, or pushed out of the way—nature red in tooth and claw. All that. We are about to learn better, and we will be lucky if we learn in time.

Getting along in nature is an art, not a combat by brute force. It is more like a great, complicated game of skill.

Altruism is one of the strange biological facts of life, puzzling the world of biology ever since Darwin. How can one explain the survival of any species in which certain members must, as a matter of routine, and under what appear to be genetic instructions, sacrifice their own lives in the interests of the group? At first glance, the theory of natural selection would seem to mandate the permanent elimination of any creature behaving this way.

Altruism is, on its face, a paradox, but it is by no means an exceptional form of behavior. It is extremely interesting to biologists, but not because it is freakish or anomalous. In most of the social species of animals altruism is essential for continuation of the species, and it exists as an everyday aspect of living. It is perhaps not so much an everyday aspect of human behavior, and there is no way of proving or disproving a genetic basis for its display when it does occur. Sociobiologists—E. O. Wilson, for example—believe that human altruism is genetically governed and exists throughout our species, whether or not in latent or suppressed form. Others, the antisociobiology faction, do not believe there is any evidence for altruistic genes at all, and attribute behavior of this kind solely to cultural influences. They do not, of course, deny the existence of human altruism; they simply deny that it is an inheritable characteristic. For all I know, either side of the argument could be right, but I would insert a footnote here with the reservation that human culture itself is not all that nonbiological a phenomenon. We may not be inheriting genes for individual items of cultural behavior, but surely we are dominated by genes for language, hence for culture itself, whatever its manifestations.

Altruism remains a puzzle, but an even deeper scientific quandary is posed by the pervasive existence of cooperative behavior, all through nature. To explain this we cannot fall back on totting up genes and doing arithmetic to estimate the evolutionary advantages to kinships. And yet it is there, and has been since the beginnings of life. The biosphere, for all its wild complexity, seems to rely more on symbiotic arrangements than we used to believe, and there is a generally amiable aspect to nature that needs more acknowledgment than we have tended to give it in the past.

We are not bound by our genes to behave as we do. Most other creatures—not all, surely, but most of them—do not have the option of introducing new programs for their survival, at will. They behave as they do, cooperate as they generally tend to cooperate, in accordance with rigid genetic specifications. It may be, probably in fact is, that we are similarly instructed, but only in very general terms, with options for changing our minds whenever we feel like it. Our options, and our risks of folly, are made more complicated by the possession of language. Using language makes it easy for us to talk ourselves out of cooperating, but the very changeability of our collective minds gives us a chance at survival. We can always, even at the last ditch, change the way we behave, to each other and to the rest of the living world. Since the time ahead cannot any longer be counted as infinite time, and since we tend to keep talking by our very nature, perhaps we still have time to mend our ways.

There are two immense threats hanging over the world ecosystem. Both of them are of our doing, and if they are to be removed we—humankind—will have to do the removing.

The first is the damage to the earth we have already begun to inflict by our incessant demands for more and more energy. Although, as I said earlier, we have not yet changed the earth's climate, it is a certainty that we will do so sometime within the next two centuries, probably sooner rather than later. We are not only interfering with the balance of constituents in the atmosphere, placing more carbon dioxide there than has ever existed before by the way we burn fossil fuels and wood, risking several degrees of increase in the mean temperature of the whole planet. We are also risking a significant depletion of the thin layer of ozone in the outer atmosphere, principally by the nitrogen oxides associated with pollution. It is a telling example of the way we think about global problems that we always talk of the ozone layer as our own personal protection against human skin cancer, as if nothing else mattered. The ecological outcome of a significant depletion of the ozonosphere would matter considerably more. A 50 percent increase in the ultraviolet band would increase the amount of UV-B at the higher energy end of the band by a factor of about fifty times. The energy of these wavelengths would have highly destructive effects on plant leaves, oceanic plankton, and the immune systems of many mammals, and could ultimately blind most terrestrial animals.

We ought to be learning much more than we know about the day-to-day life of the earth, in order to catch a clearer glimpse of the hazards ahead. One way to begin learning would be to make better use of the technologies already at hand for the world's space programs. Somewhere on the list of NASA's projects for the future is the so-called Global Habitability program, a venture designed to make a close-up, detailed, deeply reductionist study of the anatomy, physiology, and pathology of the whole earth. The tools possessed by NASA for this kind of close, year-round scrutiny are flabbergasting, and better ones are still to come if the research program can be adequately funded. Already, instruments in space can make quantitative records of the concentrations of chlorophyll in the sea—and by inference the density of life; the acre-by-acre distribution of forests, fields, farms, deserts, and human living quarters everywhere on earth; the seasonal movements of icepacks at the poles and the distribu tion and depth of snowfalls; the chemical elements in the outer and inner atmosphere; and the upwelling and downwelling regions of the waters of the earth. It is possible now to begin *monitoring* the planet, spotting early on the evidences of trouble ahead for all ecosystems and species, including ourselves.

The Global Habitability program could become an example of international science at its most useful and productive, if it could only be got under way. Right now, the chances of getting it set at the high priority it needs for funding seem slim. It has the disadvantage of offering only long-term benefits, which means political trouble at the outset. It is no quick fix. It is research for the decades ahead, not just the next few years. And it cannot be done on the cheap, which means wrangles over the budget in and out of Congress. And finally, it will require a full-time, steady collaborative effort by scientists from many different disciplines in science and engineering, and from virtually every country on the face of the earth, which means international politics at its most difficult. But it ought to be launched, and soon, no matter what the difficulties, for it would be a piece of science in aid of the most interesting object in the known universe, and the loveliest by far.

I said a moment ago that there were two great threats to the planet's viability as a coherent ecosystem. The second one is not a long-term one. It hangs over the earth today, and will worsen every day henceforth. It is thermonuclear warfare.

It is customary to estimate the danger of this new military technology in terms of the human lives that are placed at risk. We read that in the event of a full-scale exchange in the northern hemisphere, involving something like 5,000 megatons of explosives, perhaps 1 billion lives would be lost outright from blast and heat and another 1.5 billion would die in the early weeks or months of the aftermath. With more limited exchanges, say 500 megatons, the human deaths could be correspondingly reduced. We even hear arguments these days over the acceptable number of millions of deaths that either side could afford in a limited war without risking the loss of society itself, as though the only issue at stake were human survival.

But a lot of other things would happen in a thermonuclear war, more than the general public is aware of or informed about. What we call nature is itself intimately involved in the problem.

According to a study by a committee of biologists and climatologists for the Conference on the Long-Term Biological Consequences of Nuclear War, there are several probable events that will occur. Assuming that most or all of the detonations take place at ground level, the amount of dust and soot exploded into the atmosphere may darken the underlying earth over the entire Northern Hemisphere for a period of several months up to one year. The sunlight might be 99 percent excluded, and the surface temperatures in continental interiors would fall abruptly to below $-40°C$, effectively killing most plants and all forests. In the tropical zones, the loss of forests could destroy a majority of the planet's species. The photosynthetic and other planktonic organisms in the upper layers of the oceans would be killed, and the foundation of most marine food chains eliminated.

The new and extensive temperature gradients between the oceans and land masses will bring about unprecedented storms at all coastal areas, with destruction of many shallow-water ecosystems.

Radioactive fallout in areas downwind from the fireballs is estimated to expose 5 million square kilometers to 1,000 rads or more, most of this within forty-eight hours. This exposure is much higher than in any previous scenario, and is enough to kill most vertebrates and almost all forms of plant life in the affected area, including the conifers that make up the forests in the cooler regions of the Northern Hemisphere.

Later on, months after the event, things will get worse. The ozonosphere will be gone, or nearly gone, and the planet will then be exposed to the full, lethal energy of ultraviolet radiation as soon as the dust and soot have cleared away. It was only because of the protective action of ozone that complex, multicellular organisms were able to gain a foothold in life a billion years ago, and most of these creatures are still as vulnerable as ever to ultraviolet light.

The Southern Hemisphere will be less affected, assuming that the nuclear exchange is confined to the north, but extensive damage is still inevitable throughout the globe, most of it due to chilling.

Bacterial species are less vulnerable to radioactivity and cold than are the higher organisms, but many species in the soil will be lost in the initial heat of fireballs or in the later firestorms and wildfires covering huge areas.

It is not known how many forms of life would be permanently lost. After a period of years, some of the surviving species might reestablish themselves and set up new ecosystems, but there is no way of predicting which ones, or what sorts of systems, beyond the certainty that everything would be changed.

In such an event, the question of the survival of human beings becomes almost a trivial one. To be sure, some might get through, even live on, but under conditions infinitely more hostile to humans than those that existed 1 or 2 million years ago when our species first made its appearance. Civilization, and the memory of culture, would be gone forever. Given the kinds of brains possessed by our species, and the gift of memory, all that might be left to the scattered survivors, staring around, would be the sense of guilt, for having done such damage to so lovely a creature, and a poor heritage for a poor beginning.

Sermon at
St. John the Divine,
June 1984

*I*n ancient Rome, the doctors who traveled with the armies and looked after illnesses and injuries were called the immunes. These professionals were immune from public service in the expectation that they would pay strict and exclusive attention to their medical duties. In warfare ever since, doctors have gone about their business as though immune in this antique sense: they are not supposed to carry arms or to get shot at, and they have had very little say, if any, in the technology of war itself.

From time to time in recent years, the views of noncombatant physicians have been solicited and listened to by war departments, but only for limited and restricted topics. The decision to give up biological warfare was perhaps influenced by doctor-advisers, but more because of the wild impracticality of the technology and its likelihood to affect one's own troops than because of any humanitarian considerations. No general staff has as yet taken advice from their medical professionals to give up certain weapons simply because they kill too many people.

Now things have changed. The doctors of the world can no longer be professionally immune, exempt, expected to come on the battlefield after the injuries have been inflicted and do their best to fix things up. They know some things for sure about the new weapons.

Thermonuclear warfare is not just a technical problem for military specialists. It carries lethal menace to human civilization and to the human species itself. It is above all a medical problem for the political personages and their advisers who are responsible for running the governments of the earth, and also for a smaller, more exclusive group who must surely be looking carefully and thoughtfully at the problem. These are the world's armed services, the soldiery, the professional class responsible for the *profession of arms*. They are, by and large, intelligent people, educated in the science and technology of warfare, trained necessarily, by the very nature of their craft and occupation, to think well ahead of the rest of us about large-scale death. The peoples they are commissioned to serve cannot survive—on any side— the use of thermonuclear weapons. It is no longer a straightforward military game with armies measuring degrees of risk to themselves; it is human society itself that has been stationed, unwillingly, in the front lines. The politicians can always find expedients in policy and new rhetorical ways to waffle; they can always postpone decisions, hope for the best, maybe serve out their terms. The military caste is in a different situation. Neither they, nor their calling, nor their long traditions of devotion to their countries can conceivably survive the outright destruction of their societies. What lies ahead for these professionals, if even the neatest and cleanest of nuclear weapons are launched from either side, is not warfare in any familiar sense of the term. It cannot be regarded as organized combat, or technological maneuvering, or anything remotely resembling the old clash of arms. It will be something entirely new, beyond any professional adroitness in defense, beyond mending at the end. Once begun, there will be no pieces to pick up, no social system to regroup or reorganize, nothing at all to command.

But now military policy and, along with it, national policy itself are being driven by a peculiar branch of science and technology. The defense establishment's laboratories are no longer commanded to go to work and produce the specified weapons requested by policymakers; instead, the technicians produce whatever pops into their minds and then present their brand-new advances to the statesmen, changing the whole game at every turn. The politicians are thus compelled these days to tag along after the technicians, unable to make up their minds about either diplomacy or arms-control negotiations because of never knowing what will happen next month, or next week—when some technician in a nuclear arms laboratory will suddenly say, in effect, "Hold it, we're onto something new."

The latest advance is the ballistic missile with a smart war-head, which carries radar and a computer allowing it to change course during its final descent through the atmosphere, hitting its target within an estimated radius of meters instead of the kilometers of the old-fashioned ICBMs. Some of the military analysts have seized on this advance in order to defend the whole concept of nuclear warfare. One eminence in the field suggested, in an article in *The New York Review of Books*, that it should remove all our anxieties about the killing of civilians in a nuclear war. In his view, the next great war can be fought in the roman-tic style of the wars before the eighteenth century, with the outcome determined by the skill and courage of individual swordsmen. One side or the other, maybe both, can now take out, with what is called surgical precision, the weapons and the commanding officers of the other side, leaving all of the cities and noncombatant civilians undamaged. All of this, of course, on the unlikely assumption that everyone will use only the smallest bombs, tiny ones like the Hiroshima bomb, and that all of the opponent's weapons and command-control headquarters will be located at a safe distance from any towns or cities. There is a dreamlike quality in this kind of military planning, unlike anything in real life and totally unlike anything in the history of warfare and its record of blundering and folly. But there it is, a wish for nuclear warfare.

And now the same dreamy, heavy-lidded ivory tower scientists at work on their marvelously accurate missiles are also at work on nuclear defense, with all sorts of possibilities on their minds: laser beams and particle beams to zap all intruding missiles at

the point of launch or at the moment of reentry, anti–hydrogen
bomb hydrogen bombs, cities shielded by fiberglass, whole populations transported instantly to safety under the hill, countries
shielded by hope, by flags, by tears, by any old idea, by reassur-
ing strings of words.

We need a freeze, all right, but it must be a mutual freeze on
this kind of science. As a professional, I am not one to forbid
any avenues of research inquiry. But this, I think, is not real
science in my view. It has nothing whatever to do with a comprehension of nature; it is not an inquiry into nature. Its only possible outcome will be the destruction of nature itself. It should be
brought to a stop, by both sides before it gets totally out of
hand.

The military people, here and abroad, have two practical,
down-to-earth medical problems to worry about, and now they
must worry about them not only for the troops under their command but also, to a degree unique and singular in modern warfare, for the noncombatant populations they are supposed to be
defending—not just the inhabitants of the cities; the people at
large, everywhere. Moreover, although I doubt that military
handbooks have much to say about the matter, they must now
be worrying, at least late at night, about life in general, the
conjoined life of the planet. Jonathan Schell, in *The Fate of the
Earth*, writes at length and eloquently about this latter concern,
infuriating *The Wall Street Journal* and a great many practical
men of the world who do not like being told that the earth is,
as it plainly is, an immense, fragile, intricately interconnected
sort of organism, maybe best looked at as a stupendous embryo,
and capable, like all young organisms, of dying.

Two scientific discoveries dealing with the probable effects of
nuclear war upon the earth's climate and the life of the planet
have been reported, but have received only transient notice by
the country's newspapers and almost none at all by television.

The first discovery, widely known within the scientific community of climatologists, geophysicists, and biologists here and
abroad, has been confirmed in principle by counterpart scientists
in the Soviet Union. Computer models have demonstrated that

a nuclear war involving the exchange of less than one-third of the total Russian and American bombs will produce a dense cloud of dust and soot from ignited cities and forests changing the climate of the entire Northern Hemisphere, shifting it abruptly from its present seasonal state to a long, sunless, frozen night. This will be followed after some months by a settling of the nuclear soot and dust, followed then by a new malignant kind of sunlight with all of its ultraviolet band, capable of blinding most terrestrial animals, no longer shielded from the earth by the ozonosphere. In the same research, new calculations of the extent and intensity of radioactive fallout predict the exposure of large land areas to much more intense levels of radiation than previously were expected. The report is referred to as TTAPS, an acronym derived from the investigators' names: Turco, Toon, Ackerman, Pollack, and Sagan.

The second piece of work, by Paul Ehrlich and nineteen other distinguished biologists, demonstrates that the prediction of such a soot cloud means nothing less than the extinction of much of the earth's biosphere, very possibly involving the Southern Hemisphere as well as the Northern.

Taken together, the two papers ought to change everything in the world about the prospect of thermonuclear warfare. They have received already a careful and critical review by scientists representing the disciplines concerned here and abroad, and there appears to be a more or less general concurrence with the technical details as well as with the conclusions drawn. There are some dissenters, but the dissent seems to be mostly over matters of detail and over the problem of absolute certainty in this kind of research. In the view of some referees, the TTAPS report may actually be understating the climatologic damage implied by its data. It is a new world, I should think, and a world demanding a new kind of diplomacy and a new logic.

Up to now, the international community of statesmen, diplomats, and military analysts have tended to regard the prospect of nuclear war as a problem only for the adversaries in possession of the weapons. Arms control, and the endless feckless negotiations aimed at the reduction of nuclear explosives, have been viewed as the responsibility, even the prerogative, of those few nations in actual confrontation. Now all that is changed. There is no nation on earth free of the risk of destruction if any two countries, or groups of countries, embark upon a nuclear exchange. If the Soviet Union and the United States, and their respective allies in the Warsaw Pact and NATO, begin to launch their missiles beyond a still-undetermined and ambiguous minimum, neutral states like Sweden and Switzerland are in for the same long-term effects, and the same slow death, as the actual participants. If the predictions are correct, Australia and New Zealand, Brazil and the Middle East have almost as much to worry about as Germany if a full-scale exchange were to take place far to the north. If any single country were to launch a full-scale attack on any other single country, and then if nothing else happened, no retaliation, the attacking country might itself be extinguished in the months following, along with the rest of the Northern Hemisphere countries.

Up to now, the risks of this kind of war have conventionally been calculated by the numbers of dead human beings on either side at the end of the battle, armies and noncombatants together. The terms "acceptable" and "unacceptable," signifying so-and-so many million human casualties, have been used for making cool judgments about the need for new and more accurate weapon systems. From now on, things are different.

Something else will have happened at the same time, in which human beings *ought* to feel the same stake as in the loss of their own lives. The continuing existence and buildup of nuclear weapons, the contemplated proliferation of such weapons in other nations now lacking them, and the stalled, postponed, and failed efforts to get rid of these endangerments to the planet's very life, including our own, seem to me now a different order of problems from what they seemed a short while ago. It is no longer a political matter, to be left to the wisdom and foresight of a few statesmen and a few military authorities in a few states. It is a global dilemma, involving all of humankind.

I hope now that the international community of scientists in all countries will look closely at the data and conclusions reached so far, and will challenge and extend the studies in whatever ways they can think of, and will advise their governments accordingly and insistently. And I hope that the journalists of the world will find ways to inform the world citizenry at large, in detail, and over and over again, about the risks that lie ahead.

The most beautiful object I have ever seen in a photograph, in all my life, is the planet Earth seen from the distance of the moon, hanging there in space and obviously alive. Although it seems at first glance to be made up of innumerable separate species of living things, on closer examination every one of its working parts, including us, is interdependently connected to all the other working parts. It is, to put it one way, the only truly closed ecosystem any of us know about. To put it another way, it is an organism and it came into being about four and one-half billion years ago. It came alive with a life of its own, I shall guess, 3.7 billion years ago today, and I wish it a happy birthday and a long life ahead, for our children and their grandchildren and theirs and theirs.

I have a high regard for our species, for all its newness and immaturity as a member of the biosphere. As evolutionary time is measured, we only arrived here a few moments ago and we have a lot of growing up to do. If we succeed, we could become a sort of collective mind for the earth, the *thought* of the earth. At the moment, for all our juvenility as a species, we are surely the brightest and brainiest of the earth's working parts. I trust us to have the will to keep going, and to maintain as best we can the life of the planet. For these reasons, I take these scientific reports not only as a warning, but also, if widely enough known and acknowledged in time, as items of extraordinary good news. For I believe that humanity as a whole, having learned the facts of the matter, will know what must be done about nuclear weapons.

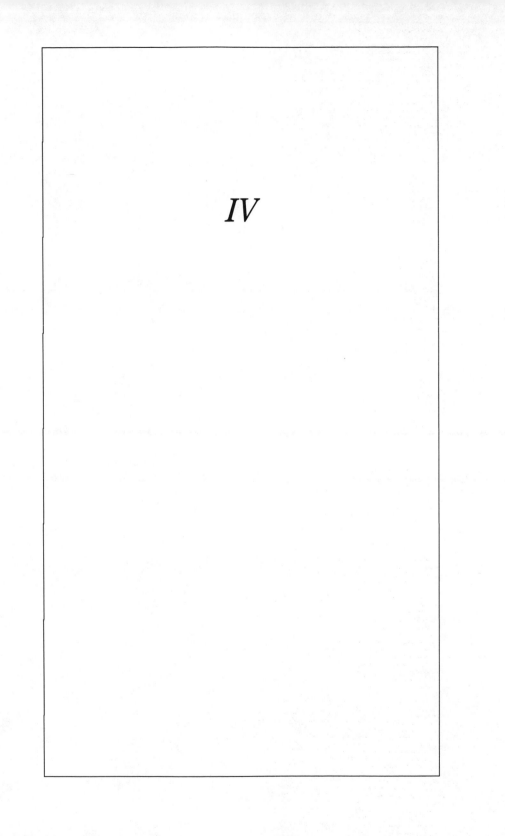

IV

Cooperation

*I*t would be only fair for me to acknowledge in advance that I intend to press a particular point of view, conceivably so idiosyncratic as to qualify as a personal bias, in this lecture and, I hope, in the two to follow. I shall be as deliberately selective as I can in the examples used to make my case, and when I run up against items of evidence contrary to my point of view I shall either argue them away or, if that is impossible, simply dismiss them. I have just one theme, and I intend to let nothing get in its way. Much of the material I shall be dealing with is what people in my own field refer to as anecdotal, a generally pejorative word when applied to scientific evidence. The only term harsher than "anecdotal" in scientific criticism is "trivial," and I shall be risking that as well. Furthermore, I plan to deal with technical matters considerably beyond my own training and professional competence, exactly as though I know what I am talking about, relying upon secondhand information from investigators in other disciplines whose work I trust and whose evidence happens to fit my bias. In short, I shall be haranguing nearly as much of the time as I hope to be informing.

My argument, stated generally and briefly, is that the driving force in nature, on this kind of planet with this sort of biosphere, is cooperation. In the competition for survival and success in evolution, natural selection tends, in the long run, to pick as real winners the individuals, and then the species, whose genes provide the most inventive and effective ways of getting along. The most inventive and novel of all schemes in nature, and perhaps the most significant in determining the great landmark events in evolution, is symbiosis, which is simply cooperative behavior carried to its extreme. But something vaguely resembling symbiosis, less committed and more ephemeral, a sort of *wish* to join up, pervades the biosphere.

This notion, I hasten to insert defensively, applies mostly for the very long run. In the short term, viewing what we see around us on any particular day, it seems to fly in the teeth— one might say the teeth and claws—of the evidence.

But still, there are small pieces of evidence for the notion at work even in short spans of time. Herewith, to make this point, my first anecdote.

The story begins with some experiments at the University of Buffalo in the late 1960s with various strains of amoebae. A fastidious and meticulous technique had been developed for removing the nucleus from one amoeba and then replacing it with the nucleus of another, opening all sorts of new and exciting approaches for cell biology and genetics. It had been learned that a nucleus could be successfully transplanted among the amoebae of any particular strain, functioning as nicely as the one that had been sucked out, really an astonishing event considering the regulatory and reproductive complexities in the transactions between a nucleus and all the other components of its cell. Even more surprising, however, was the observation that this could not be done when the exchange was made between amoebae of two related but slightly different genetic lines. The nuclear transplant, in such cases, behaved rather like a miniature tissue graft between unrelated mice, with graft rejection as the inevitable outcome. Whether the alien nuclei rejected the host cells, or the cells the nuclei, remains unsettled; in any case the transplanted chimeras promptly died. At hand was a new technique for examining the intimacies of the relation between a nucleus and its cell, and the genetic factors involved in the relationship.

Then, as often happens in the course of a particularly engrossing venture in basic research, a catastrophe occurred. Dr. N. K. Jeon, who was then doing the work at the University of Tennessee, discovered that his amoebae were reproducing at a sharply decreased rate, looked sick indeed, and were beginning to die off. On a closer look he found that the whole line had been invaded by rod-shaped bacteria, with as many as 150,000 microbes inside each cell.

In other circumstances, with less future research at stake, the amoebae would ordinarily have been discarded and a new line started up, but these were valuable animals, on their way to revealing the self-marking genes of cellular nuclei. Therefore, Jeon nursed them along in hopes of recovery from the infection. Parenthetically, although invasion and infection are the correct terms for the situation at this point, there must have been an earlier, unrecognized stage at which things would have seemed the other way round. Amoebae have achieved their own success in evolution by virtue of two conspicuous attributes, motility and phagocytosis. Bacteria cannot get inside such a cell unless they are, sequentially, chased, captured, and engulfed. Early on, at the beginning of the affair, it would have looked more as if the bacterial population had been invaded by amoebae. But then, having made their meal, the predators were taken over by their prey, and the latter, instead of being digested, began to colonize their hosts. Something like this happens from time to time in the infections of higher organisms, including ourselves; typhoid fever and tuberculosis are instances where the host's own phagocytic defense cells turn into the primary sites for the multiplication of the invading bacteria.

Jeon was able to keep his cultures going, although with difficulty, and after many months of laborious subculturing and nursing, the amoebae regained their health and resumed reproducing at their normal rate. They were not, however, free of the bacteria by any means; each cell still contained no less than 50,000 microbes. At this stage, the easy explanation would have been that the amoebae were somehow acclimated to their residents, or that the bacteria had somehow lost their toxicity for their hosts. But, as it turned out, something much more subtle and profound had taken place.

The amoebae, after months of living with their bacteria inside, had become dependent upon them and unable to live without them. When treated with appropriate antibiotics, the bacteria died off but then, soon thereafter, the amoebae died as well. The same thing happened when the cultures were heated to a temperature calculated to kill the bacteria but not the amoebae: both parties promptly died.

The nuclei of the adapted, infected amoebae had undergone, during the process of adaptation, a fundamental change. When they were transplanted to an identical line that had never been infected, the nuclei were now lethal for normal amoebae. In contrast, when the nucleus of an infected amoeba was replaced by the nucleus from another infected cell, the transplant was uniformly successful.

Thus two important events transpired in the interaction between the host cell and pathogen, both of which appear to involve a genetic adaptation enabling the infected amoebae to survive and prosper. The pathogen was assigned the totally new role of an indispensable organelle, and the amoeba nucleus changed its label from self to a discernibly different self, probably something like "self + x." Of these events, the most interesting and allusive is the change from pathogenicity to indispensability on the part of the bacteria. The mechanism underlying this remains beyond clarification at present, since the bacteria themselves cannot yet be cultivated outside of the living amoebae. It is unlikely, in my view, that the changed situation results from any alteration in the bacteria or their properties, and indeed there is no evidence for this in Jeon's experiments. What seems to have happened is that the infected cells not only learned to resist the toxic action of the bacteria, but also, perhaps just because of their continuing existence in the cytoplasm, learned to make use of them, and finally to depend on them for life itself.

In more recent work, Lorch and Jeon have extended the experiments and discovered that the adaptive change in the amoeba nucleus, as well as the shift of the interaction from parasitism to symbiosis, takes place with remarkable speed. The whole transformation, representing on its face a small-scale model in evolution, can be brought about within as short a time as six weeks and after no more than sixteen generations of infected amoebae.

This seemingly freakish event can be taken as a sort of biological parable, reminding us of what must have been one of the three or four most crucial turning points in the evolution of life on earth—the formation of complex, nucleated cells like ours, known now as eukaryotic cells, from their bacterial predecessors. The event probably occurred sometime around 1 billion years ago.

Before this could possibly have happened, a great many other things had to take place to set the biological stage. Indeed, the longest stretch of evolutionary time, fully three-fourths of the planet's life, seems to have been preoccupied by single cells trying to organize themselves into a workable system.

To go back to the earliest stage, the very beginning of life, requires more than a leap of the imagination and a reliance on paleontological data which become sparser the further back you look. It also takes a measure of modesty and humility for our own species. Perched as we are, or think we are, at the very pinnacle of biological progress, we must face the fact that our origin was humble indeed. Our interest in genealogy tends to run back only for the number of generations needed to prove that we came from lines of kings. A few of us, specialized for the inquiry, can take us back through more remote ancestors, beetle-browed, small-brained, hairy, even tailed, swinging through trees. But the fact of the genealogical matter, like it or not, is that the original ancestor of humanity was a single cell, almost undoubtedly with a form and functions resembling those of one or another of today's common bacteria. It may even be necessary to go further back, assigning our lineage to a single strand of RNA assembled by accident in a lightning storm nearly 4 billion years ago.

It is the purest kind of guesswork, but there are some solid facts to use for guessing. The oldest fossil record indicating any kind of life on earth is in rocks assuredly dated at least 3.7 billion years old. The planet itself was formed around 4.7 billion or so years ago, so it took, roughly speaking, a billion years for life to emerge in a sufficiently complex form to leave microscopic traces of itself. From that time on, the bacteria multiplied and spread themselves across the face of the earth, and they were for all those years the only creatures alive. It was not until a relatively short time ago, only 600 million years back, that macroscopic forms appeared with large- and solid-enough bodies to leave fossils visible to the naked eye. From there on it is relatively easy going, and most of today's impressive body of evolutionary theory is based on what lies in the paleontological record of just the past half-billion years.

But the one thing we do know for sure about our bacterial ancestors is that they learned, very early on, to live in communities. The solid evidence for this is in the stromatolites, complex mounds of laminated rock in which the various layers were both constructed and occupied by colonies of various species of prokaryotes. The fossilized stromatolites are remarkably similar to today's algal mats, structures to be found in coastal marshes in many parts of the earth, laminated in much the same fashion. These contain live colonies resembling the ancient fossil forms—anaerobes living off sulfur, others using carbon dioxide and producing methane, other species living off the methane and producing new organic nutrients for the next layer up, blue-green algae (now called cyanobacteria) living off the sun and producing oxygen, plus countless spirochetes and other microbes, all housed in what can only be viewed as an immense household.

Most of the microorganisms living together in this style—and this is true also for countless other bacterial species apparently living free in the earth's soil—cannot be isolated from each other and cultivated in conventional bacteriological media. Because of this, very little is known about their metabolic functions or nutritional requirements, beyond the conspicuous fact that they live together and cannot live apart.

There seems to be no argument about two outstanding features of the earliest forms of bacteria. They must all have been anaerobes, in view of the firm geological evidence that the earth's early atmosphere contained no oxygen, or at most only faint traces. And the first ones must have lived off the ground, literally, for lack of any other organic matter around except themselves. There is no special trick involved in either of these features, of course. Anaerobes abound everywhere today, or at least in every place where they can avoid exposure to oxygen (a lethal gas for most such creatures), deep in the mud of swamps or in the shelter of other bacteria capable of consuming any oxygen in the vicinity. Moreover, there are still vast numbers of them able to make a good living with nothing but sulfur or other inorganic material for food and energy.

It has until recently seemed likely that these ancient bacteria represent, as an anaerobic class, the direct lineal descendants of the very first cell, or cluster of cells, that emerged on the planet between 3.5 and 4 billion years ago. Now, however, based on the powerful techniques for the chemical analyses of RNA made possible by molecular genetics, it appears that one large subgroup, the so-called archaebacteria, are the strongest candidates for deep antiquity. Among the archaebacteria are the methanogens, the anaerobes that still make swampfire by converting water and carbon dioxide to ordinary illuminating gas. The RNA of these creatures is similar to that of several even weirder bacteria—the halophiles, which flourish in saturated salt brine; the thermoplasmas, delicate wall-less mycoplasmas that have been found living at high temperatures in smoldering piles of coal tailings; and a third group of conventional-looking bacteria, the thermoacidophiles, which can only exist in the extremely hot acid waters of hot springs. These microorganisms are the pet project of Carl Woese at the University of Illinois; he believes that they are the oldest of living things and perhaps the real parents of us all. It is an attractive idea if you are looking about for biochemical analogies connecting us to them. One species of archaebacteria, called *sulfolobus,* grown from a hot spring, has been found to have DNA similar to ours in one significant respect: the genes

are interrupted by what are called "introns," or intervening sequences of DNA. The function of these strings is not understood,
but they have been regarded up to now as one of the unique
features of modern nucleated cells and are never present in ordinary bacteria.

But it is the ability of the archaebacteria to live in such hostile
environments that makes them most attractive as candidates for
primogeniture among the earth's creatures, especially their capacity to live and grow at very high temperatures. If, as is commonly assumed, the first cells on earth had to be pieced together
by a series of random accidents involving chemical materials already at hand in the earth's waters, the statistical probability of
such an event occurring within even a billion years can be seen
as vanishingly small. To achieve, all at once, a flawless code for
the complex information system of DNA and RNA, which would
later become the universal code for all the earth's creatures 4
billion years later, has seemed so entirely improbable that Francis Crick and Leslie Orgel regard it as outright impossible; they
propose seriously that bacteria must have been planted here by
extraterrestrial visitors long ago, bucking the problem of life's
origin to other scientists, someplace else in space. Their theory,
called "Directed Panspermia," is no more improbable than the
present notion, and also, of course, no less.

But if there were enough heat, the random events needed for
the ten-strike would have moved along much more rapidly. It is
for this reason that the "hot-smoker" bacteria have become, in
the past few years, the most interesting and also controversial
topic for contemporary biology. These are methane-producing
bacteria claimed to exist in extremely hot water, under very high
pressure, in vents in the earth's crust at the bottom of deep-sea
trenches. They were isolated by Baross and his associates from
water spuming out of sulfur cones at temperatures in excess of
300°C.—water which could only remain water at such temperatures because of the tremendous pressure of the overlying
ocean. Now it is claimed that they can actually be cultivated and
grown in the laboratory under similar conditions. Trent, at the
Scripps Oceanographic Institution in La Jolla, California, dis-

agrees with the claim and believes the presumed bacteria are really artifacts, and hence the controversy. I cannot think of a more important issue in all of science. If Baross is right, an entirely new and plausible scenario for the origin of life is there for everyone to understand and think about. And I might add that even if he is wrong, the thought of random events occurring at the enormous speed induced by such heat, and in water, plus the knowledge that many of the archaebacteria which indisputably exist today live comfortably at lower temperatures but still close to boiling, will keep the scenario alive as a hypothesis for years to come. We may, after all, have been born in fire.

Whatever their origin, the hardest of all subsequent tasks for primordial bacteria was to learn to survive the first appearance of oxygen in the atmosphere. There is no precise date for the event, but it may have begun around 3 billion years ago when new mutants capable of photosynthesis appeared—the counterparts of today's cyanobacteria. Thereafter, oxygen steadily increased in the atmosphere, reaching a level around 1 percent 2 or so billion years ago. At this stage, any bacteria exposed to the atmosphere must have developed biochemical mechanisms to protect themselves, and soon thereafter, new species, the aerobes, emerged with the capacity to make efficient *use* of oxygen for their energy needs.

The whole earth then began, in a sense, to breathe. The photosynthetic organisms, employing nothing more than sunlight, water, and carbon dioxide, gradually filled the air to its present level of 20 percent oxygen, making carbohydrate in the process. The respiring bacteria, now surrounded by an abundance of organic matter, used the new excess of oxygen for burning their food.

Now the stage was set for the greatest of all evolutionary steps, the invention of nucleated cells. It could not have happened without the existence of the two major participants, the photosynthetic cyanobacteria and the respiring aerobic bacteria. Somehow or other, these must have become incorporated as symbionts in the flesh of some larger, still unknown prokaryote, turning, respectively, into the organelles known as chloroplasts in green plant cells and mitochondria in all nucleated cells. The first host cell may have been one of the mycoplasmas, expansile organisms lacking cell walls. Or it could have been a bacterium which had previously lost its own rigid wall, like one of today's L-forms. Ever since, these organelles have maintained the arrangement, reproducing themselves on their own, independent of the cell nucleus, and preserving most (but not all) of their own DNA and RNA. The unmistakable marks of their bacterial ancestry are still there in the most modern plant and animal cells. They are permanent lodgers, absolutely essential for the life of what we like to call higher forms, stunning examples of the power and stability of symbiosis in the advance of evolution.

An advance it was, and is. It is fashionable in some quarters these days to deny the notion of progress in nature, and to assert that evolution does not result in anything like a hierarchical increase in the complexity or profundity of living things. This notion is perhaps defensible if you are only glancing back a short distance, the 600 years, say, that preoccupied the attention of biologists in the nineteenth and early twentieth centuries. But when you look all the way back, at the time when nothing existed but bacteria, and then at the evolutionary moment when these organisms joined up to form nucleated cells, and then at the fantastic events since that moment, it is hard to resist the thought that progress is the way of evolution, and the way of the world. Unsteady progress, perhaps, with everything jerked back into partial extinction every 25 million years or so by cyclic collisions with asteroids, but progress ahead nonetheless.

But maybe not so much as to sweep us off our feet. During that long period when bacteria were the sole occupants of the earth, a great many beautiful experiments in biochemistry and genetics must have been going on, and biological novelties must have been turning up on every side, some workable, some not. Long before the invention of endosymbiosis and the eukaryotic cell, the bacteria were engaged in interliving in microscopic eco-systems, living off each other, establishing themselves in ecological niches, signaling to each other and working out message systems and information receptors within their communities.

Some of the chemical signals that we have been regarding as our own sophisticated hormones, specially made for our higher tasks in life, were already being synthesized by the simplest bacteria long before, perhaps billions of years before we or anything like us came on the scene—a molecule resembling insulin, for instance, with properties indistinguishable from those of insulin, or peptides similar to our own growth hormone. Others are still being identified and characterized. It may be that much of what we have been calling endocrine secretions, maintaining the integrity of our own huge assemblages of nucleated cells, are the direct inheritances of molecules employed originally by the bacteria for the regulation and maintenance of their early communities.

We have risen, to be sure, and there is no question about our eminence in nature at the moment. We possess brains, opposing thumbs, and all sorts of other appurtenances beyond any imaginable reach of our bacterial ancestors. But we carry traces of that ancestry nonetheless, and it will be good for our minds, and for the way we look at our place in nature, to learn more about them. It is also nice to know that there are things some of them could accomplish easily that are still beyond our reach— navigation by internal compass, for example. We know that pigeons and honey bees can find their way around by small magnets embedded in specialized organs, and perhaps someday structures like these will be found to exist in the human brain. If so, and even so, we are not demonstrably good at using them, and they are by no means a modern invention. Certain species of ancient anaerobic bacteria, in need of reliable ways of telling up from down in order to swim down where they flourish best, deep in the mud, are equipped with tiny chains of crystalline magnetite, running the length of their bodies like a string of beads. When placed in a magnetic field, with one pole representing the North Pole, the bacteria immediately swim in that direction; when the poles are switched, the creatures turn like dancers and swim in the opposite direction. In the Northern Hemisphere, these bacteria habitually swim toward the North Pole; in New Zealand, they swim south. In equatorial Brazil, they have difficulty deciding which way to go. We humans may be the cleverest of all animals, as we tell ourselves, but we haven't really run away with the game, not yet anyway.

But symbiosis is far and away the smartest trick of all. The chloroplasts and mitochondria are perhaps the most spectacular examples, but there are many others, equally important for the survival and evolutionary progress of the partners involved. Lynn Margulis, who contributed much to the endosymbiosis concept, is now pursuing the idea that prokaryotic spirochetes may have attached themselves to some of the earliest nucleated cells and, later on, evolved into the cilia of modern cells. The same idea entails the view that the basal bodies of nucleated cells, now known to be essentially involved in the process of mitotic division, came from the same symbiotic association with spirochetes.

Although still unproven, the idea is not as wild as it first sounds. Spirochetes are nearly as famous for symbiosis as they are for producing syphilis. The intestinal tracts of certain termites contain motile protozoans which owe their motility entirely to the masses of spirochetes attached to their surfaces, beating in absolute synchrony to move them from place to place.

The termite is itself a sort of living paradigm of symbiosis. The whole termite nest is a model of cooperative behavior, seen close up. The hundreds of thousands of separate insects behave rather like the individual cells of a huge organism, almost like a brain on countless legs. Each of the individuals appears to carry a genetic blueprint for the intricate architecture of the termitarium, made of columns and arches at its base, chambers specially designed for the queens, and the entire construction air-conditioned and humidity-controlled. But then, each individual termite is itself an assemblage, a sort of functional chimera. The insect lives on wood, but does not possess its own digestive system for converting cellulose to usable carbohydrate. This is the function of the motile protozoans, single-celled eukaryotes which inhabit its intestine and are passed on from one generation to the next by feeding. The protozoans are, in certain species, unable to move about by themselves to ingest the fragments of wood eaten by the termite. They accomplish their movement thanks to the spirochetes attached to their surfaces. And there is more to come. Inside each protozoan, embedded in neat layers just beneath the creature's surface, are numerous bacteria; these are the ultimate symbionts, contributing the enzymes needed for digesting the wood eaten by the termite, located by the spirochetes, swallowed by the protozoans, and now awaiting conversion to sugar by the bacteria. It is as eminently successful a committee as can be found anywhere in biology.

The habit of interliving hangs on in the bacteria of the earth, and life would be meager without them. Although they have gained themselves a reputation as pathogens, this is based on the aberrant behavior of a tiny minority of bacteria. Indeed, pathogenicity may represent a mistake, a misinterpretation of signals between the organisms involved, rather like the initial events in Jeon's infected amoebae. Much more often, the bacteria-host arrangement is one of partnership. The rhizobial bacteria seem to be invading and infecting the root hairs of clover and soybeans, forming large colonies within the root tissues, for all the world like an overwhelming infection. But the outcome is the fixation of nitrogen for the life of the plant itself, and, not incidentally, for the subsequent enrichment of the soil in which the plants are growing. The multitudinous bacteria in the foregut of all ruminant mammals are essential symbionts, playing the same role in the breakdown of cellulose as their counterparts in the termite. We would have neither cattle nor milk without the steady work of their microbial lodgers.

I do not know why certain fish in the sea need to have luminous tissues around their eyes, but they do, maybe as an aid in locating prey or attracting mates. Whatever the purpose, luminescence is a species-specific and evidently inheritable characteristic of these fish, and yet it is entirely the work of symbiotic bacteria colonized within the tissues.

Cockroaches and other insects contain whole organs among their tissues that are made up of nothing but bacteria, closely packed together. What they are doing there is entirely unknown, except that they are essential and are passed along from generation to generation. If they are eliminated by antibiotic treatment, the insects slowly languish and die.

Much more research needs to be done on the general phenomenon of symbiosis, and it will be difficult work indeed. A major
technical difficulty, always in the way but especially troublesome
for field biologists, is to recognize stable symbiotic partnerships
when they do indeed exist. The partners are often difficult to
distinguish from each other without ultrastructural analysis, and
usually impossible to separate from each other in the living state
for biochemical analysis. My guess, based on the guesses made
by many ecologists, is that the phenomenon is much more common than is generally believed, perhaps even a commonplace.
Even as we see it today, unusual, exceptional, even exotic, it
has the look of something of extreme importance. As Margulis
wrote a couple of years ago, "Symbiosis has affected the course
of evolution as profoundly as biparental sex has. Both entail the
formation of new individuals that carry genes from more than a
single parent. . . . The genes of symbiotic partners are in close
proximity; natural selection acts upon them as a unit."

I take the view that the successful and persistent existence of
symbiosis, quite aside from the question as to whether it is
relatively uncommon or relatively all-over-the-place, suggests
the underlying existence of a general tendency toward cooperative behavior in nature. It is simply not true that "nice guys
finish last"; rather, nice guys last the longest. The great ecolo-
gist Evelyn Hutchinson pointed out, some years ago, that complex natural communities do not follow what *ought* to be the
mathematical rule implicit in the term "survival of the fittest." If
that term ever meant "winner take all," some one species, any
predator with advantages in growth rate and numbers of surviving offspring, should always predominate, outclassing and out-
growing all the others. Out in the field, this does not happen.
On the contrary, stable ecosystems composed of hundreds of
different species tend to remain stable and in balance. Only rarely
does a single species exclude all the others, and when this happens that species, of course, crashes. The Crown of Thorns
starfish once frightened the biologists who study aquatic reef
systems when it seemed that this predator was spreading like a
contagion and on the verge of eliminating all coral reefs in the

Pacific. I had a laboratory at Woods Hole at the time, and I remember vividly the agitation in all the corridors: the reefs were vanishing, the atolls were threatened, the life of the ocean itself was at stake; *do* something, find a starfish poison, get an army of scuba divers off to Australia. Then something happened, beyond scientific analysis, called Nature Taking Her Course, and the problem simply went away. The Crown of Thorns is still there, browsing away, perhaps now being browsed at by other smaller creatures, but anyway under restraint, doing its level best to cooperate.

Human beings would do well to look very closely at this situation, and at others like it, and to mobilize the resources of marine science to examine the matter. We may well be in a comparable dilemma, and we had better learn soon how the Crown of Thorns got out of its fix. I believe we already have grounds for some collective hunches about our own fix, based on the mere handful of years we've been in existence as a social species.

It is easy enough to account for the existence of genuine altruism in nature. Among the most intensely social species it is a commonplace event, part of any day's work, for one individual member to sacrifice his life in aid of the community. The honey bees, for instance, who must eviscerate themselves because of their barbed stings whenever they attack intruders in defense of the hive, are bona fide altruists in biological terms. It is, in a perfectly straightforward sense, self-preservation, since what is being preserved are the genes of the particular bees that do the defending and stinging. In biology, the preservation and persistence of one's genes represent reproductive success, and therefore evolutionary success. Behavior of this kind would naturally be naturally selected in the course of Darwinian evolution.

Very likely, something like true altruism exists among other species as well, including our own, although it is more difficult to identify it as genetically driven behavior when it does occur, even when it takes place within human family groups. Also, to be sure, it is not a regular or a predictable phenomenon and no one, especially in today's social setting, would claim that siblings or parents go about sacrificing themselves for each other every day of their lives. J. S. Haldane's summary of the mathematical justification "I would give up my life for two brothers or eight cousins" is really a speculative abstraction laying out what Haldane would do if he were exclusively under the control of Haldane's genes rather than Haldane's culture, which is a doubtful assumption.

Nevertheless, theoretically anyway, altruism makes good biological sense for any species intent on the preservation of its line, provided it doesn't get out of hand. But what about cooperation, which is rather different from altruism? The stakes are not the same, the genes involved between cooperating partners are not identical, not necessarily even related, and nobody is required to give up his life. Is there any way to account for pure cooperation, as opposed to pure altruism, in terms acceptable to current evolutionary theory? Are there circumstances in which cooperative behavior can be expected to confer an advantage upon an individual within a generally uncooperative, egoistic species, or upon a species within a world of generally uncooperative other species?

Perhaps so. Robert Axelrod of the University of Michigan has recently demonstrated, at any rate, that computers say that it is so. If you are curious to know how it was that a great many soldiers managed to stay alive during the worst phases of the carnage imposed by the trench-warfare system in World War I, cooperation is unquestionably the right answer. A social historian, Tony Ashworth, wrote a whole book on this matter, entitled *Trench Warfare, 1914–1918: The Live and Let Live System.* At the beginning of the war, German and Allied soldiers faced each other from trenches a few hundred yards apart and simply let fly all available explosives at each other, day and night, resulting in a huge number of deaths on both sides. As the war proceeded, however, a quite different strategy began to emerge, by a sort of unstated but mutual agreement between the units on both sides, beyond the control of the generals and their staffs far behind the lines. In essence, what had started out as full-scale destruction, motivated by a kill-or-be-killed policy, turned into something more like live-and-let-live. It was like a game.

The players learned to recognize when food and other supplies were being brought up behind the enemy's trench lines, and firing ceased during some of those periods. Whenever a battalion defected from the arrangement and bombarded the other side's supply line, it found its own supply lines under a much more intense fire than normal, as retaliation. Whenever the troops involved found themselves in prolonged combat, facing each other over periods of weeks or months, the firing tended to quiet down to an almost symbolic level. It was distressing to the high command on both sides; soldiers were court-martialed, whole battalions were disciplined, but the accommodation and reciprocity went on. According to Ashworth, the tactics of trench warfare were incorporated into what he termed "ritualized aggression . . . , a ceremony where antagonists participated in regular, reciprocal discharges of bombs which symbolized . . . sentiments of fellow-feelings and beliefs that the enemy was a fellow-sufferer." There are, of course, no biological conclusions to be drawn from this record. The soldiers knew what they were doing, and were simply applying common sense in order to keep the numbers of dead at the lowest possible level.

But, in retrospect, there are biological lessons indeed. Axelrod and Hamilton set up an elaborate computer game, based on game theory, to determine what happens when two or more totally egoistic participants confront each other over and over again for long and indefinite periods of time in competition for essential resources. The players have two choices on each encounter: to cooperate, in which case both benefit to a limited extent; or to defect, in which case one receives a much larger benefit at the other's expense, unless both players defect, which yields no benefit to either side. It is a modification of the old game-theory paradox called the Prisoners' Dilemma.

Axelrod enlisted the help of some seventy-five experts in mathematical game theory, computer intelligence, and evolutionary biology, and set up a series of computer tournaments to learn whether there are any particular strategies that win over others in the test of cooperation versus defection. A large number of programs were submitted, some highly complex, some devious and clever, some treacherous, some outright exploitative and consistently aggressive. They were pitted against each other in a long series of round-robin games, also tested against a program in which all decisions were made by random choices. Each program was run against all the others, including a clone of itself, 200 times, and then the whole tournament was repeated five times in a row.

The results are contained in an extraordinary book by Axelrod, entitled *The Evolution of Cooperation*. The hands-down winner, and the simplest strategy of all, was the program sent in by Professor Anatol Rapaport, a philosopher and psychologist at the University of Toronto and a longtime puzzler over the Prisoners' Dilemma game. Rapaport's program is called, in conventional computer jargon, TIT FOR TAT, and it stipulates merely the following strategy: Cooperate on your first move, and thereafter repeat whatever move the other player makes. If he defects, you defect; if he cooperates, you cooperate.

As it turns out, TIT FOR TAT is not a strategy for any quick victory, it is no way of clobbering opponents, it always loses some but also always wins a few more, and when played enough times against enough other players, even the most clever and devious ones, it always wins in the long run. It is what Axelrod calls, for want of a better technical term, a "nice" strategy, but not "too nice." It cooperates with cooperation, retaliates against betrayal, remembers, forgives, and can be trusted.

It is also a strategy that will inevitably, and with mathematical certainty, spread through any community of players using other strategies. A cluster of TIT FOR TAT strategists cannot be invaded by other, hostile or aggressive players. Once established within a sea of competitors—provided the game goes on for an indefinite period of time—it emerges as the only game in town. It does not require intelligence or anything like intuition—indeed, there is something basically counterintuitive about TIT FOR TAT when viewed as a strategy for the short run. It fits nicely with the kind of behavior we might have predicted for the numberless microorganisms cooperating in a modern algal mat or a Pre-Cambrian stromatolite, and it would fit as well for the life in a tropical rain forest. The computer game persuades us that it is the kind of biological logic that one might expect to emerge by natural selection in the course of evolution.

I never expected news like this from computer science. As an outsider, a nonplayer, I always thought computer games were contests between combatants bent on eating each other up or blowing each other up, exactly like what one sees in the behavior of human committees and, most of all, in the behavior of modern nation-states. Now I am all for the computers, and I hope the word gets around quickly.

Communication

Other creatures, most conspicuously from our point of view
the social insects, live together in dense communities in such
interdependence that it is hard to imagine the existence of anything like an individual. They are arranged in swarms by various
genetic manipulations. They emerge in foreordained castles,
some serving as soldiers for defending the anthill or beehive;
some as workers, bringing in twigs of exactly the right size
needed for whatever the stage of construction of the nest; some
as the food-gatherers tugging along the dead moth toward the
hill; some solely as reproductive units for the replication of the
community; even some specialized for ventilating and cleaning
the nest and disposing of the dead. Automatons, we call them,
tiny genetic machines with no options for behavior, doing precisely what their genes instruct them to do, generation after
mindless generation. They communicate with each other by
chemical signals, unambiguous molecules left behind on the
trail to signify all sorts of news items of interest to insects:
the dead moth is on the other side of the hill behind this rock,
the intruders are approaching from that direction, the queen
is upstairs and asking after you, that sort of news. Bees, the

earliest and greatest of all geometricians, dance in darkness to tell where the sun is and where it will be in exactly twenty minutes.

We, of course, are different. We make up our minds about the world as individuals, we look around at the world and plan our next move, we remember what happened last week when we made a mistake and got in trouble, and we keep records for longer memories, even several generations back. Also we possess what we call consciousness, awareness, which most of us regard as a uniquely human gift: we can even think ahead to dying, and we cannot imagine an insect, much less a wolf or a dolphin or even a whale, doing *that*. So we are different. And marvelously higher.

Nonetheless, we are a social species. We gather in communities far denser and more complex than any termite nest or beehive, and we depend much more on each other for individual survival than any troupe of army ants. We are compulsively, biologically, obsessively social. And we are the way we are because of language.

Of all the acts of cooperative behavior to be observed anywhere in nature, I can think of nothing to match, for the free exchange of assets and the achievement of equity and balance in the trade, human language. When we speak to each other, it is not like the social insects laying out chemical trails; it contains the two most characteristic and accommodating of all human traits, ambiguity and amiability. Almost every message in human communication can be taken in two or more ways. There are choices to be made all over the place, in the sending of messages and in their reception. We are, in this respect, unlike the ants and bees. We are obliged to listen more carefully, to edit whatever we hear, and to recognize uncertainty when we hear it, or read it.

Another difference is that the communication systems of animals much older than our species are fixed in place and unchangeable. Our system, language, is just at its beginning, only a few thousand years old, still experimental and flexible. We can change it whenever we feel like, and have been doing so right along. How many of us can speak Chaucerian English, or Anglo-Saxon, or Indo-European, or Hittite? Or read them?

But it is still a genetically determined gift, no doubt about it. We speak and write and listen because we have genes for language. Without such genes, we might still be the smartest creatures on the block, able to make tools and outthink any other animal in combat, even able to think and plan ahead, but we would not be human.

It is not clear whether the gift of language turned up because of a mutation, suddenly transforming us from one kind of species into a distinctly different creature by the installation of brand-new centers for language, or whether you get language automatically, with a big-enough brain. It could be either way. We could be human because of specialized centers designed for grammar, as songbirds evolved identifiable neuron clusters on one side of their brains for generating birdsong; or we could be generating grammar, and transforming it, simply because we acquired brains huge enough to do this sort of thing.

It is not a trivial question, and sooner or later we can hope to settle it since it is a question open not only to speculation but also to scientific inquiry. Did we become human because of acquiring the property of language, or did we gain language as the result of acquiring human brains?

It is not a question for the birds, although the birds have hypotheses for us if we like them. A song sparrow sings his elaborate song, stereotyped in its general message but ornamented by himself alone, because he possesses a large, sharply delineated cluster of neurons in the left fronto-temporal cortex. If he hears the typical song of his species early in his life, as a nestling, he will remember it ten months later and sing it accurately, with a few modulations of his own. If he is kept from hearing it in his childhood, he will never learn it. If he is exposed to the song of another species, a swamp sparrow, say, he will sing a strange medley of song-sparrow and swamp-sparrow sounds on maturity. If deafened as a nestling, he will sing nothing beyond a kind of buzz. The cells responsible for the song of a canary are typical, conventional-looking neurons, easily recog-

nized in stained sections of the brain, but between mating sea-
sons they die away and vanish. Then, with the next spring, they
reappear as full-fledged brain cells, synaptic connections and all,
and the song center is back in place. This is a nice piece of
news: brain cells can die off and then can regenerate themselves.
Until Notebohm and his associates at Rockefeller University
learned this from their studies of birdsong, a few years ago, we
had taken it for granted that brain cells could not regenerate;
once dead they were gone for good. We knew that the olfactory
receptor cells out in the open nose can regenerate all the time,
every three weeks or so, but that seemed an extraordinary exception. Now, with the canary to contemplate, we can begin to
guess that maybe any part of the brain might be given the same
capacity for regeneration, if only we knew more about the regulatory mechanisms involved. I never used to like canaries much,
nor their song, but now, as a once-neurologist, I hear them
differently.

As I said a moment ago, human language is probably at its
earliest stage, just beginning to emerge and evolve as a useful
trait for our species. When we first acquired it is not known,
but we do possess something resembling a fossil record. Around
20,000 years ago, our human ancestors scratched marks on
rocks which Marshak and other scholars have interpreted as
primordial arithmetic and accountancy. The proto-Sumerian tab-
lets from the fourth millennium B.C. contain clear records of
calculation, mostly concerned with barley measures, based on
sequences of sixty rather than the tens used in our system. We
can make good guesses about the origins, even the dates of
origin, of many of the words in our language today, thanks to
the scholarship of comparative philologists over the past two
centuries. We can even see now, from this distance, how we
made most of our metaphors from their words: true, for instance, from their *deru* for "tree"; world, for instance, from their
wiros for "man." Someone remarked, just a few years back, that
we are much more knowledgeable than any preceding generation. T. S. Eliot, when told this, remarked yes, but they are
what we know.

Human culture has evolved in a manner somewhat similar to biological evolution. If, in studying the development and variation of several different languages, you can find consistent similarities between certain words of those languages, you are permitted to deduce that another word, parental to all the rest, existed at some time in the past in an earlier language. It is the same technique as is now used by today's molecular biologists for tracking back to the origin of today's genes. The ordinary bacteria, for instance, have sequences in their DNA which are significantly different from those in the DNA of modern, nucleated cells, and also just as different from those in the so-called archaebacteria—the ancient, anaerobic methane-producing bacteria of very long ago. The molecular geneticists viewed this problem in the same way that the early-nineteenth-century philologists viewed the similarities of Greek to Latin and all the Celtic, Germanic, and Slavic languages, and of these to Sanskrit and before, and deduced the existence of an original language, Indo-European, antedating all the rest. For the molecular geneticists, a theoretical species called the "U-bacteria," speaking in ancient but still recognizable biochemical words, serves the same function as in Indo-European philology.

I have invented my own taxonomy of language, not for the purpose of classifying different national or ethnic modes of speech but simply to divide up language into the several different purposes for which it is used. My classification reduces language to four principal categories.

The first category is small talk. This is human speech without any underlying meaning beyond the simple message that there is a human being present, breathing, and at hand. We use this communication mainly at social gatherings. The modern cocktail party is the best place to listen for it, but I have no doubt that an equivalent mode of speech has existed as long as society. The words do not matter, nor is grammar or syntax involved. The only messages conveyed are indications of presence: I am here. For this purpose, ritual phrases describing the weather, the route just taken by automobile, the flowers on the piano, the piano itself, suffice.

This usage of language is also employed by humans for the declaration of territory, for the competitive acquisition of space within a room filled with other humans, and, on occasion, for the beginning of courtship.

The second category is proper, meaningful language. This is where real cooperation begins. This is the marketplace of humanity. Thoughts are thought up, packaged, unwrapped and packaged again, worried over, then finally put out, but never for sale, only given away. It is the strangest market in the world. Nothing is sold, no payment ever asked or received; everything is given away in expectation of something to be given in return, but without any assurance.

Speech is, at its best, the free exchange of thought. When it is working well, one human being can tell another everything that has just happened as the result of the explosive firing of one hundred billion nerve cells in his brain, wired together by a trillion or more synaptic connections. No amount of probing with electrodes inserted into the substance of the brain, no array of electroencephalographic tracings, can come close to telling you what the brain is up to, while a simple declarative sentence can sometimes tell you everything. Sometimes a phrase will do to describe what human beings in general are like, and even how they look at themselves. There is an ancient Chinese phrase, dating back millennia, which is still used to say that someone is in a great hurry, in too much of a hurry. It is *zou-ma guan-hua*; *zou* means "traveling," *ma* means "horse," *guan* is "looking at," *hua* is "flowers." The whole phrase means riding on horseback while looking, or trying to look, at the flowers. Precipitously, as we might say, meaning to look about while going over a cliff.

Language is more than a system of signals and directions; it is a mechanism for describing what is going on within a mind. Most often it is used for pointing out the connections between one thing and another, seemingly different thing. It relies heavily on metaphor, using words and images fitted together in such a way that something quite new and different is revealed. The fourteenth-century nun Julian of Norwich, wanting to describe how the world seemed to her, wrote this: "And He showed me a little thing the size of a hazelnut in the palm of my hand, and it was round as a ball. I looked thereupon with the eye of my understanding and thought: What may this be? And it was answered generally thus: it is all that is made." The language is itself very simple and the sentences seem and sound crystal-clear, but this short section of Julian's *Relevations* has reverberated in all its ambiguity down through six centuries, serving even the purposes of today's advanced cosmological physicists.

Nobody knows when human language first began, or how. It is anyone's guess. One sort of guess is that the brain center for generating speech turned up as a mutation, and was then selected because of its obvious Darwinian advantage and thereafter spread through the species. This strikes me as highly improbable. The earliest human beings whose brains resemble those of *Homo sapiens* did not live together in one interbreeding community capable of passing special genes along. They were continents apart, with no possible way for that first mutant to get around from one community to another like Johnny Appleseed, spreading the new kind of seed. Mutations come at random, affecting individual members of a species long before they can affect the whole species, and language is so complex and intricate a mechanism that to suppose a mutation in the brain for just that one function would presuppose an extremely rare event. It could not possibly have occurred repeatedly and independently in one human population after another, always the same mutation.

More likely, the gift of speech came along with the gradual evolution of the human brain itself, needing nothing more than the great size and intricate connections between cells to become a possibility, along with the lucky accident of having the right kind of oral cavity, tongue, palate, and larynx to permit vocalization. Sooner or later, given a big-enough brain, any creature would begin communicating in this way. Even the tongue and larynx are not essential, not even the ears for listening to speech. Children who are born deaf and mute can learn sign language just as quickly as normal children, and they go through developmental stages of language acquisition that are remarkably similar to those among speaking children. The first sign language in young deaf-mute children exhibits the same kind of baby talk, with the same mistakes in syntax and the same early difficulty in distinguishing between the terms for "you" and "me" that are the normal experience of speaking children. Mature sign language is very much like mature speech; there are a great many ways of signing the same meaning, ambiguity can be expressed easily, and the signs for ordinary language can even be transformed to something like poetry.

If language is a universal gift of human beings, part of what comes along from possessing an ordinary human brain, and if we therefore do not have to puzzle over the near-impossibility of a solitary mutant starting up the business somewhere on earth, we are still left with the problem of how language itself started. Did everyone just begin talking, all at once, as soon as our brains reached a certain size? Did grammar and syntax, flexible sentence structure, and a real vocabulary simply pop into our heads all at once? Or did they come gradually, in stages? As we look around at the several thousand different languages now in use around the world, there is little to help us here at hand. Noam Chomsky, who was the first to propose a coherent theory for the biological origin of language, has not yet succeeded in identifying the necessary underlying structures required for a complete theory of universal language, and perhaps the various languages now in existence are fundamentally too different from each other. Many linguists maintain that there is no such thing as a primitive tongue; all languages are adapted to the environments in which the speakers live, but all are equally complex and subtle. Indeed, some of the languages spoken by people in remote, primitive communities are more complicated and certainly much more highly inflected than English, and the oldest of all existing languages, Chinese, is the least inflected and in some ways the simplest and clearest. Benjamin Lee Whorf pointed out that not only do different environments result in the emergence of totally different languages; the languages themselves impose totally different ways of looking at the world. The Hopi language has no words or idiomatic constructions for what we call time; the world simply *is*, it does not change over time in any causal way. Whorf says the language is better than English for mapping the terrain of twentieth-century physics. The Algonkian languages are incredibly subtle in the terms available for intricate social relations. Whorf wrote: "A fair realization of the incredible degree of the diversity of linguistic system that ranges over the globe leaves one with an inescapable feeling that the human spirit is inconceivably old; that the few thousand years of history covered by our written records are no more than the thickness of a pencil mark on the scale that measures our past experience on this planet . . . , that the race has only played a little with a few of the linguistic formulations and views of nature bequeathed from an inexpressibly longer past." This, wrote Whorf, need not be "discouraging to science but should, rather, foster that humility which accompanies the true scientific spirit and thus forbid that arrogance of the mind which hinders real scientific curiosity and detachment."

There simply *must* be a universal scheme for making language, even though the languages that are put together by that scheme are wildly different in detail from each other, even though very few if any details of real importance can be identified as universal to all languages. Whether Chomsky is right, that the same deep structures in the mind are the source of generative transformational grammar in whatever language, cannot yet be established, but the failure thus far is the fault of the still-primitive state of neurobiology. Perhaps indeed we all have centers for grammar in our brains, analogous to but far more complex than the known centers for speech itself, perhaps broadly analogous to the centers for birdsong in Marler's and Notebohm's birds.

But even when we reach this level of understanding, and have some neuroscience to account for the existence of the physiology of language, we will still be stuck with the other question: how did it start, and who started it? Did committees get themselves appointed in the earliest tribes of hunters and gatherers, chosen from the wisest and most experienced elders, to figure out better ways of communication beyond simply pointing at things and howling or growling? Did the committees then make up lists of words and lay out rules for stringing them together to make sense? Or was something more spontaneous at work?

Ma, mamma, pa, and *papa* are as close to linguistic universals as you can get. These words, or words sounding like them, exist in many of the world's languages, and they were probably first spoken by very young children. That is, of course, a guess.

The word "pupil," with the two meanings of the pupil of the eye and a small child, may have acquired both meanings in the same way from children. The Indo-European root was *pap*, the word for the nipple or the breast, which with some kind of logic turned into terms for very small children: *pupus* and *pupa* in Latin, then *pupillae*, then pupil. Every language derived from Indo-European has the same connection, and for the same reason: when someone looks very closely into someone else's eye, he sees the reflection of himself, or part of himself. But why call that part of the eye a pupil? The same duplication, using identical terms for the pupil of the eye and a child, occurs in totally unrelated languages, including Swahili, Lapp, Chinese, and Samoan. Who would most likely have made such a connection, and decided then to use the same word for a child and the center of the eye? Most likely, I should think, a child. Who else but a child would go around peering into someone else's eye and seeing there the reflection of a child, and then naming that part of the eye a pupil? Surely not, I should think, any of the members of a committee of tribal elders charged with piecing together a language; it would never cross their minds. The pupil-eye connection must have turned up first in children's talk.

Which brings me to Derek Bickerton and his theory to explain the origin of Creole languages. Bickerton, a professor of linguistics at the University of Hawaii, has spent much of his career on the study of Hawaiian Creole, a language that developed sometime after 1880, when the Hawaiian Islands were opened up for sugar plantations and needed lots of imported labor quickly. The new arrivals, joining the existing communities of English-speakers and native Hawaiians, came from China, Korea, Japan, the Philippines, Puerto Rico, and the United States, each group, with its own language, unable to communicate with any of the rest. As always happens in such a circumstance, a common pidgin speech quickly emerged, not really a language, lacking most of the essential elements of grammar, more a crude system for pointing and naming items and giving simple directions. Most of the words were hybrids made by combining words from the collected languages, or imitations of English words. Pidgin is itself such a word, a mispronunciation of "business English."

At some period between 1880 and 1910, Hawaiian Creole emerged as the universal speech of the islands, enabling all the younger workers to speak together. The Creole was qualitatively different from Pidgin, a genuine formed language with its own tight rules for sentence structure, grammar, and word order, its own variants of articles and prepositions, its own inflections and indicators of tense and gender—in short, a brand-new human speech.

According to Bickerton, who had the opportunity to interview some of the first settlers, when Hawaiian Creole first appeared it could neither be spoken nor understood by the original adult workers. It was the language of the first generation of children, and must have been constructed, in almost its entirety, by those children.

Bickerton asserts that Hawaiian Creole is a unique language, fundamentally different from the tongues spoken by the parents of the language makers. He claims, in addition, that this Creole resembles, in important linguistic details, other Creoles that have appeared at other times after similar language catastrophes in other parts of the world—in the Seychelles, for example. Other scholars, especially the linguistic specialists known professionally as Creolists, have disagreed with him on these claims, and argue that Hawaiian Creole contains linguistic features similar enough to the parent languages to allow for the possibility of borrowing some grammar. But they do not argue, so far as I know, with his central point: that Hawaiian Creole could not possibly have become the common speech of the plantations as the result of any participation by the adults; it was neither taught nor learned by the adults, and it must therefore have been invented by children.

Bickerton has another point to make. Hawaiian Creole has certain features which make it technically similar to the word arrangements and syntax of young children being raised everywhere in the world. Thus, in his view, it recapitulates a stage partway along in the acquisition of language by children. From this, he argues that it can be taken as hard evidence for the existence of what he calls a universal "bioprogram" for language acquisition and language itself, which I take to mean a center or centers within the human brain responsible for both the generation and learning of grammar.

If Bickerton is right, or even partly right, his observations place children in a new role as indispensable participants—prime movers indeed—in the evolution of human culture. Everyone knows that young children are spectacularly skilled at acquiring new languages, the younger the better. They are positively brilliant when compared to adolescents, and most adults are out of the game altogether. All that a young child needs to pick up a new language is to be placed at close quarters with other children speaking that language. It is, literally, child's play.

I can imagine a time, long ago, I can't guess how many thousands of years ago, when there were only a few human beings of our sort, with our kinds of brains, scattered in small clusters around the earth: the early tool-makers, cave dwellers, hunters and gatherers, some living in isolated family households, others beginning to form groups, bands, tribes. No language yet, lots of hoots, whistles, warning cries, grunts. Then some words, the names of animals, trees, fish, birds, water, death maybe. A good many of these must have been imitative, like the Indo-European root *ul*, which originally meant simply to howl, later becoming *uurvalon* in Germanic, then *ūle* in Old English, and finally, in English, "owl," the bird. Along the way, it drifted into Latin as *ululare*, meaning "to howl," into Middle Dutch as *hūlen*, finally into English as the words with precisely the same meaning as in the original Indo-European, from *ul* to "howl" and "ululation" in heaven knows how many generations.

Let us assume, then, the spontaneous development of some sort of lexicon, antedating any sort of language, within every early human community. Things in the environment would have agreed-upon names. Very likely people would have names as well. Human speech, at that stage, would be limited to signals and markers, rather like a modern pidgin speech, but it would still be a poor way to transmit human thought from one mind to another. At the same time, every human brain would be capable of language, although language did not yet exist. How then did it begin?

I suggest that it began in the children, and it probably began when the earliest settlements, or the earliest nomadic tribes, reached a sufficient density of population so that there were plenty of very young children in close contact with each other, a critical mass of children, playing together all day long. They would already have learned the names of things and people from their elders, and all that remained for them to do was to string the words together so that they made sense. For this, they used the language centers in their brains, assembling grammar and constructing syntax, maybe at the outset, in much the same fashion that birds compose birdsong. To set off the explosion, and get it right, you would need a dense mass of children, a critical mass, *at* each other all day long for a long time.

When it first happened, it must have come as an overwhelming surprise to the adults. I can imagine the scene, the tribe gathered together in a communal compound of some sort, ready to make plans for the next hunt or the next move, or just trying to discuss the day's food supply as best they can in grunts and monosyllables, mildly irritated by the rising voices of the children playing together in a nearby clearing. The children have been noisier than ever in recent weeks, especially the three- and four-year-olds. Now they begin to make a clamor never heard before, a tumult of sounds, words tumbling over words, the newest and wildest of all the human sounds ever made, rising in volume and intensity, exultant, and all of it totally incomprehensible to the adults holding their meeting. In that moment, human culture was away and running.

Awhile back I ventured on a classification of human language but got only as far as small talk and ordinary language and then tripped over the children. I have two more to add to the taxonomy.

The third category is an entirely new form of communication, assembled from bits and pieces of logic over the past several centuries and now beginning to turn into the first and thus far the only genuinely universal human language. Parts of it can be spoken, all of it is written, and it bears no relation whatever to the parent languages of those who use it. It is the language of mathematics.

If you want to explain to someone else how the universe operates at its deepest-known levels, all the way from occurrences in the distant cosmos to events within the nucleus of an atom, you cannot do so with any clarity or even any real meaning using any language but mathematics. The twentieth-century world of quantum mechanics is the strangest of all worlds, incomprehensible to most of us and getting stranger all the time. Yet it is indisputably the real world. It seems to change everything we have always taken as reality into unreality, threatening our fixed ideas about time, space, and causality, leaving us in a new kind of uncertainty and bewilderment perhaps never before experienced by our species. A wave is a wave, but it is also a particle, depending on how we choose to observe it.

None of this can be expressed or understood or explored in any language other than mathematics, nor can the real substance of it be translated into English. Written rapidly in chalk on a blackboard, the ideas can be taken in and then debated by an audience whose members need share no other common language—English, French, Germans, Italians, Russians, the lot. You can't buy a cup of coffee in mathematics, but you can explain, by all accounts, nearly everything. There are some theoretical physicists here and abroad who are so confident of the recent progress in their field that they believe all of the essential nontrivial questions about the material universe can be satisfactorily answered by a grand unifying theory, perhaps before the end of this century. This may indeed happen, but if it does, all the answers will only be expressible in the language of mathematics. For most of us, lacking mastery of that tongue, the world will then be an even stranger place than ever. The philosophers, carrying their traditional responsibility for explaining matters like this to the rest of us, will be out of business unless they are schooled in the highest mathematics beforehand; perhaps one of them will learn to translate, or at least to interpret, the real substance. Parenthetically, this might turn out to be the most pressing of all reasons to begin transforming our educational system at the secondary and university level. We will need a much larger population of young mathematics speakers (or writers), not just for the future of technology and engineering or science in general, but for catching at least a glimpse of how the world works in its new reality.

I have a fourth category of language in mind, but only to mention in closing. It is poetry, which I take to be an extension of communication beyond all the normal usages of language itself. It is not as indecipherable to the untutored mind, but in some ways it is just as different from ordinary language as is higher mathematics. At its best, it is as hard to explain as music, and I have no intention of coming close to either of those formidable problems.

Only to say this much: children have had a hand in it, and childhood may be the single period in a human life when poetry begins to take hold. Without that long, puzzling period of immaturity which characterizes our species, poetry might never have entered human culture and would surely never have swept along the evolution of culture as it has since as far back as anyone can remember.

Where the children come in is at the beginning, setting in place this ungovernable and wonderful aspect of the human mind. They do so with their nursery rhymes, as close to music as speech can come. Nobody claims that nursery rhymes were written by children, but they were surely ornamented and converted into a dance of language by children, and very young children at that. They have been passed along from generation to generation by the children themselves, no doubt with the help of mothers but mostly child to child. They exist in all languages, always with the same beat, rhythm, and rhyme. Iona and Peter Opie, in their introduction to *The Oxford Dictionary of Nursery Rhymes*, say that they are "the best known of verses in the world, not at all the doggerel they are popularly believed to be." Robert Graves wrote, "The best of the older ones are nearer to poetry than the greater part of *The Oxford Book of English Verse*." G. K. Chesterton asserted that "Over the hills and far away" is one of the most beautiful lines in all English poetry, and Swift, Burns, Tennyson, Stevenson, and Henley all swiped the line for their own purposes.

Some of the rhymes seem to have been passed around among the world's children as though they were a separate tribe from the rest of us, speaking a common language none of the rest of us know anything about. The counting rhymes used for choosing the players in a children's game are just one example among many:

> *Eeny, meeny, miny, mo*
> *Barcelona, bona, stry*

is the Opies' recording of the song of Wisconsin children.

In Germany it goes:

> *Ene, tene, mone, mei*
> *Pastor, lone, bone, strei*

In Cornwall, England:

> *Ena, mena, mina, mite,*
> *Basca, lora, hora, bite.*

In New York, dating back to 1820:

> *Hana, mana, mona, mike,*
> *Barcilona, bona, strike.*

There are other older variants, in which some of the seemingly nonsense words are different, and the differences may be important.

In Edinburgh:

Inty, tinty, tethere, methera

In America, an old counting rhyme known as "Indian counting":

Een, teen, tether, fether, fip.

These are nearly the same words as those used for counting sheep by Celtic shepherds before and after the Roman occupation of Britain, and passed along by oral tradition for centuries. They may have been picked up by Welsh children 2,000 or more years ago.

Compare, for instance, the nineteenth-century American rhyme:

Een, teen, tether, fether, fip,

and one from Scotland:

Eetern, feetern, peeny, pump,

and the shepherds (numbers in ancient Northumberland):

Eeen, tean, tether, mether, pimp: 1, 2, 3, 4, 5.

The Opies conclude that the Celtic language was best preserved during the Roman occupation by people who lived in isolation, especially those needed most by the Roman garrisons, the shepherds and stock-breeders. They believe that it may have been from these adults that Welsh children picked up their numbers, and preserved them, with some modifications ever since, spreading them by their own oral tradition across Europe and then over the Atlantic.

It is another good reason for respecting children, even standing in awe of them. The long period of childhood is not just a time of fragile immaturity and vulnerability, not just a phase of development to be got through before the real show of humanity emerges onstage. It is the time when the human brain can set to work on language, on taste, on poetry and music, with centers at its disposal that may not be available later on in life. If we did not have childhood, and were able somehow to jump catlike from infancy to adulthood, I doubt very much that we would turn out human.

Connections

I make two assumptions to start with. First, the central problem that most fearfully menaces the human community for the years just ahead, despite all the heartening excitements of recent European political transformations (perhaps, indeed, the more to be worried over *because* of these transformations), is nationalism. The modern nation-state, even at its democratic very best, is an inherently unstable and unpredictable organism, and any *system* of nation-states, bound together by circumstances of trade, industry, and politics, is all the more in danger. Such a system exists in ceaseless and often random motion, comparable to other nonlinear dynamical systems in that the whole apparatus may at any time veer off into chaos, and is even in very good years subject to minor episodes of dementia when no one is looking.

My second assumption is that some powerful steadying cohesive force is needed now more than ever before in order to hold the close attention of each nation on the interests of all the neighbors, all around the world, improving the moral climate on the international scene, grasping the opportunity presented by the innate tendency of all living organisms to engage in symbiosis whenever the chance presents itself, and placing heavy bets on synergy as a natural outcome of transnational collaboration. One candidate I would put forward as such a steadying force for human society, enhancing the comity of nations, is (of course) basic science.

It may be useful to recall here that the fine old phrase "comity of nations" was designed by its etymological roots to mean something more than just an absence of rancor, or even sustained peace itself, certainly much more than international good manners, diplomatic courtesy. "Comity" comes to us from the Indo-European root *smei*, "to smile," and the phrase means, or ought to mean, nations smiling together, *co-smis*, even smiling upon each other, a state of affairs not yet attained anywhere on earth but still not beyond imagining. Anyway, I shall imagine it, and I propose that the relatively new and already spectacular phenomenon of international collaboration in science, still spontaneous and ungoverned, totally nonnationalistic in its origins and, in my view, best left that way, is a high hope for the future.

Examples of the phenomenon are all over the place, most conspicuously in any week's table of contents of the journals *Nature* and *Science*. In the October 19, 1989, issue of *Nature*, the results of the Soviet Phobos 2 exploration in the neighborhood of Mars and its strange moon are summarized; there are fourteen papers, filling almost the entire issue, all about Phobos, with a total of more than a hundred contributors. As might be expected, most of the authors are Russians, but by no means all, and the footnotes beneath the names on each of the fourteen papers demonstrate the transnationality of the whole endeavor: the home-base laboratories of the non-Soviet participants include Berlin, Pasadena, Providence, Helsinki, Tucson, Orsay, Paris, Toulouse, Greenbelt, Maryland, Los Angeles, Noordwyk, Orleans, Budapest, Sofia, Graz, Ann Arbor, Lindau, and Maynooth, Ireland.

And although the scope and magnitude of the Phobos project were unusual, the same tendency toward multiauthored papers, with contributors based in widely separated laboratories in every part of the world, is becoming more and more commonplace, in every field of science. In molecular biology, now booming along in every research university as well as a great many biotechnology industrial laboratories, another interesting feature of contemporary science can be seen. The individual laboratories have themselves become multinational centers; this is, I think, especially a feature of today's American basic science scene. From the very names of the authors, it is obvious that many American research universities, and many industrial laboratories, could not be functioning at anything like their present scientific productivity were it not for the presence of increasing numbers of young researchers from Europe and the United Kingdom, but also from Japan, China, India, Taiwan, and other Asiatic nations. Indeed, I make the guess (in confidence, of course) that for all our troubles with a foundering of the American educational system and the diversion of so many of our talented native youth from science to Wall Street, we may yet be saved for a decade or two, perhaps until we have pulled up our national socks, by the influx of the Asiatic young.

What strikes me now as most extraordinary about all this is that it works so well and seems to be so much fun for all concerned. This indeed is a surprise in itself, and one more thing to worry about. Can it always be fun for young people to be caught up in increasingly "big" science, risking near anonymity by the sheer numbers of other names alongside or in front of theirs on the same paper? Can the excitement and the fun last, with so much ambiguity about who gets which credit for what piece of the work? I trust this will work itself out over time, but only if the whole system for doing basic science on an international scale continues to expand and grow. And, in the biomedical sciences anyway, I wish at the same time for the rediscovery of what can be done in small laboratories.

In the best of worlds, the steady, healthy growth of basic science is precisely what should happen, and I hope it will. Already, the rate of sheer intellectual tumbling across all national borders is an amazement. Thus far, as we near the end of this otherwise dreadful century, the phenomenon confirms Peter Medawar's assertion in his preface to *The Limits of Science*: "Science is a great and glorious enterprise—the most successful, I argue, that human beings have ever engaged in."

One lovely thing about the enterprise, now approaching full tilt, is that the governments of the world seem hardly to have noticed that it is going on. Basic science, I mean. Not at all applied research, not for a moment the profitable technologies that emerge from science. The bureaucrats of the world know all about these matters, and the later stages of research are already tangled in the meshwork of international patents and commercial treaties, subject to obtundation by bureaucrats and regulatory civil servants at every turn. But, so far anyway, basic research is in the clear.

To be sure, basic science costs a lot of money and has to be paid for, and the major source of the support has always been government, but there has always been a difference between the mechanisms for supporting basic science and those for the development and application of technology. I like to think, as an American, that the style for the difference was in part, at least, set by the NIH, in the early 1950s, and the credit for enlighten- ment should go in large measure to Dr. James Shannon, who literally invented the notion that basic inquiry in biomedical re- search should be the responsibility of the federal government. Application and development could be left to the private sector, but the sustenance of the human capacity for making pure guesses about the workings of nature and taking chances on any good hunch, since unlikely to be taken up and bankrolled by business, should become the obligation of government. Now, in a world aloud in recriminations over the fallibility of any sort of bureaucracy, it is a historical pleasure to look back at the early years of NIH as one of our government's rare and unequivocal intellectual triumphs.

But for the sustenance of basic science to remain forever in the clear, or even within the years just ahead, will take some doing. We need constant reminding that almost all of the dazzling feats of high technology that have come on the scene in recent years, transforming our information systems, our transportation, our energy sources, to a lesser extent our agriculture, and per- haps now in its early stages our capacity to cope effectively and decisively with human disease, have dropped into our laps as the unanticipated products of basic research done years ago for reasons totally unconnected with any forethought of the technology now at hand.

A timely example of this sort of inquiry can be found in the origins of today's biotechnology, especially that branch of the endeavor called, infelicitously to my ear, genetic engineering. Already, after just a few decades, the field has moved from a state of such ignorance in the early 1940s that the term "gene" was a kind of abstraction, used in detachment from any notion of what such a structure might be made of, indeed not even a consensus that a gene possessed a structure. Then, beginning with the discovery by Avery, MacLeod, and McCarty of DNA as the structure, and the elucidation of its fine architecture by James Watson and Francis Crick, investigators all around the world settled down to play with this new thing.

"Play" is exactly the word for the activity, all the way through the next three decades, one enchanting surprise after another, finally the restriction enzymes for cutting and then splicing DNA; the insertion of genes of one organism into the genome of another, recombinant DNA; and to the astonishment of everyone, including the researchers most intimately involved in the work, a new technology, something alarmingly important, something marketable, an entire new industry in the making.

It didn't, of course, come along as easily as I've just made it sound. Looking back on it, one sees that it was a long line of extremely hard work, hotly competitive, frustrating to a lot of investigators. Nonetheless, it was the greatest fun for the imaginative winners in the games, one grand game after another. The energy which drove the hard work along was uncomplicated and irresistible: it was the urge to find out how a singularly strange and engrossing part of nature works.

If this line of work had been launched, at its outset, with an administrative prediction, a tight organization, task assignments and flow charts all over the place, aimed at developing something useful and usable, potentially profitable enough to launch a new industry, my guess is that the work would have gone nowhere. It would have been administered by centrally placed committees and commissions, all shots would have been called in advance, including the results to be obtained and the deadlines to be met. There would, because of tidy management, have been no moments of surprise, no shouts of laughter at the sheer impossibility of this unexpected result or that, no changing of minds in midtrack. And, with nationalism driving the enterprise, there would also have been, fatally, in the nature of things, confidentiality in every institution, secrecy in every laboratory, enough to stop the progress of the work at every new turn.

In real life, luckily for all concerned, there were no real secrets anywhere. To the contrary, the people engaged in the work, worried as they sometimes must have been by the compulsion to publish before the competing group in Melbourne or Cambridge or Paris or wherever, sweating out the weeks of manuscript review, nevertheless kept no real secrets even when they wished to. Indeed, it is one of the marks of a really good scientist, particularly a young one, that secrets are nearly impossible to keep. Part of the pleasure in doing science comes at the moment when the results are in and irrefutable, when one can get on the long-distance phone in San Diego and call Oxford, then rush out onto the street and tell any passerby who will stop to listen.

This is still the way the system works, but I am apprehensive for the future. Up to now, basic science has been acknowledged everywhere—or almost everywhere—as a good in itself, valuable to society as a means of finding out how the world works, thus meeting the one ineluctable need that distinguishes *Homo sapiens sapiens* from all our neighbors on the planet: an always nagging, never satiable curiosity about nature and our place in the arrangement. For this overriding reason, and as a tribute to our abiding common sense as a species, we have seen to it in recent centuries that scientific inquiry is underwritten one way or another; governments in general, whether generous or parsimonious in their support, have legitimized the pursuit of deep science. But now, with profits to be made for industries at home, and under more stringent demands by nationalistic interests, more pressures exist to generate scientific advances at home and keep them there.

The scientific community has emerged in this century as the only genuine world community that I can think of. It has had nothing to connect it to the special interests of nation-states, it carries out its work of inquiry without respect for national borders, it passes its information around as though at a great party; and because of these habits, science itself has grown and prospered. Every researcher, in whatever laboratory, depends for the work's progress on the cascades of information coming in from other laboratories everywhere, and sends out his laboratory's latest findings on whatever wind is, at the moment, blowing.

I do not believe that science can be done in any other way, and I hope and pray that the world community of basic scientists can stay free, exchanging everything they know for the pure pleasure of the exchange—*competing*, of course, but playing with delight against all odds in a huge endless game rather than in a narrow contest for new business or new weapons. For science is only at its beginning, and almost everything important lies still ahead to be learned about and comprehended. Technology is another matter, the general public's business, subject to whatever supervision and regulation the public may decide to install; science must have its own open air and its freedom to take chances.

There is no end to the problems lying ahead, waiting for solution. The technological problem of energy for the world's population matches in enormity the problem of that population itself, beyond my capacity to make guesses beyond the assertion that neither of these can be solved until we have learned new, still unguessed-at things from basic research. Somewhere on anyone's list are the puzzles raised by technologies already a part of everyday life, suddenly turning out to be doing more harm than good and needing replacement.

Making lists of basic science priorities is quite a different challenge from listing the kinds of technology we would like to have for the future. Any such lists should be made by quite different groups of experts, and for different motives. The basic science items, almost obviously, should contain at the top the deepest puzzles, the aspects of nature about which we are most embarrassed by our ignorance, needing enlightenment for the plain reason that we cannot endure the awareness of ignorance. Human beings are all right for as long as they are ignorant of ignorance; this is our normal condition. But when we know that we do not know something, we can't stand it. The technologies are no easier to arrange in priority, but at least we know here what it is we want and need, and we can have some confidence that if we are patient enough and work hard enough we can get what we want—fusion or solar energy or both, for example.

I would put agriculture at the very top of any list I might make of scientific areas ready for fundamental change. Some of the brightest of molecular biologists, casting about for new types of cells for playing with, have discovered the existence of plant cells, and are turning themselves into molecular botanists overnight. Not because they wish to grow tropical rain forests or grow gigacorn crops, but simply because plant cells are suddenly fascinating and open to intimate genetic manipulation and transformation, which the molecular biologists came late to realizing. Now they are about to be off and running with these new cells, and all sorts of things could now be lying ahead, ready to revolutionize the crops of the earth. We will soon be told that corn or wheat or soybeans can be cloned for whatever constituents we like best, grown in huge vats as single cells, and served up for breakfast, with cloned milk atop. But if these frivolities are not attained, we will at least be able to feed the famished populations of the world.

And we should be preserving the genes of plants, preserving the planet's biological diversity, as best we can. The United States has maintained a national seed bank for many years, located in refrigerated vaults in a modest three-story building in Fort Collins, Colorado. This facility, the equivalent for plant breeders of the Library of Congress, contains about a quarter of a million samples of crop seeds and their wild relatives, offering a vast treasure of plant germ plasm and an irreplaceable assortment of genes available for the breeding of new, hardier, higher-yielding, disease- and pest-resistant crops for the future needs of agriculture. It is something of a surprise, almost a shock, that this National Seed Storage Laboratory, NSSL as it is known within the tiny governmental agency responsible for its maintenance, has never been heard of by the American public at large and has to get along on a shrinking budget totally inadequate for its requirements. Still, it is indeed one of the nation's treasures, as are the equally small and inadequate seed banks stored in a few other places around the world.

In 1770, Benjamin Franklin was serving as colonial agent in England, and this wise man sent home the seeds for America's first soybeans. Some years later, Thomas Jefferson displayed a comparable prescience. Not a man given to hyperbole, Jefferson wrote, "The greatest service a citizen can do for his country is to add a new crop for his countrymen."

The value of a comprehensive seed bank is now incomparably greater than ever before because of the new methods available for the isolation, characterization, and cloning of individual genes. Whatever environmental disasters may lie ahead for mankind, including the likelihood of global warming (and the equally likely global chilling for which a new cyclic ice age must be waiting for us, perhaps within the next millennium), humankind will be needing for survival a vast and diverse population of genetically novel and adaptable plants. Moreover, and perhaps of greater importance, the very existence of seed banks should be a stimulus for expanding basic research in molecular genetics to include plant life in general, an almost guaranteed road to new astonishments awaiting the biologists of the world.

Such a resource should soon open the way to a detailed, deeply reductionist exploration of the general phenomenon of symbiosis, still a relatively undisturbed problem in basic science. Nothing, in my view, could be more important in the long run for our comprehension. After all, symbiotic partnerships are as close to implying an underlying law of nature as the rules of quantum mechanics are for physics, but we still know very little about the mechanisms involved, even less about the governing role symbiosis must have played in evolution. Plant life offers nice models for such studies, plant-insect partnerships, plant-prokaryote interdependencies, and the like. And we can hope for some handy spin-offs: a clarification of symbiotic interliving will open the way to new notions about parasitism. The accommodation of spirochetes to integration in higher cells as cilia, or in trypanosomes as flagella, or the ancient incorporation of other prokaryotes as the mitochondria and chloroplasts of all modern cells: all are now ready for detailed scrutiny. In short, there is everything to do.

Other fields of questioning are opening up to the scientific intellect every passing year. The neurobiologists are falling all over each other in their excitement over the new lines of inquiry into the brain and its doings; the notion that consciousness itself may soon become a biological problem as well as a philosophic one is no longer an embarrassing thing to talk about in public. The virologists are beginning to recognize the critical role played by their subjects in passing around genes among the world of microbes, speeding up evolution at rates unimaginable for Darwinian gradualism; the viruses, some of them anyway, may someday be regarded as a third sex, diverse not only in themselves but producing even more genetic diversity among their hosts.

And then there are the mathematicians, laying down their novel and inscrutable laws out in the wings, approaching the footlights only when it dawns on the rest of us that some of our most difficult puzzles demand for their solution the very sets of equations invented more than a half century ago by the mathematicians for their own, private entertainment. There is no field of basic science I can think of that is not dependent for its past accomplishments, even more for whatever it may hope for its future, on theoretical mathematics.

So it is appropriate to note here, not just in passing, that the world of mathematics provides the rest of us with the best and most exemplary of all models for the internationality, transnationality of science. The mathematicians have been talking to each other for centuries, over great distances, back and forth across the borders of every nation-state, giving away their information and always getting back more, *trading* human thought on a global scale. If we are looking about for new ways to enhance the apparatus of international collaboration in basic science, we would do well to examine how the mathematicians went about the business.

It may be that we should be leaving the complicated system alone, not meddling with it, allowing it to grow and change all on its own. It might even be prudent not to be discussing the matter, lest our friends in government and industry suddenly sit up in their chairs, alert at last to the possibilities for better management. Let it be, don't fix it, don't even talk about it.

I don't know. I do believe that we should at least be worrying about the system. And while I am fearful of what would happen if nationalism were to become a moving force in the organization of science, I possess a painfully acute awareness of where the money comes from to make science possible. The governments of the earth, and now some of the world's industries, are in a certain hard sense our masters, and their comprehension of what international science is up to, and how it can best do its work, is crucial. The masters need to be kept informed, and they could surely make use of sound, objective advice from the world scientific community, self-serving as that advice is bound to sound to many ears.

Part of my education in this matter was acquired during a six-year period as a member of the United States invention known as the President's Science Advisory Committee, PSAC for short, in the last years of the life of the committee before its quiet euthanasia at the hands of President Nixon. PSAC was, in its time, an exceedingly useful structure for the U.S. government. It had around twenty members, representing at that time mostly the physical sciences but with a sufficient minority of biomedical and social science types. The committee met for two days every month, with an agenda of items sent down by the top White House administrative chiefs, but with ample time for considering matters raised within the committee itself. The chairman was the White House science adviser, a presidential appointee; in years past this job had been occupied by such figures as Kistia-kowsky, Wiesner, Dubridge, Hornig, and David, people held in high regard by the American scientific community and also trusted by the politicians for their objectivity and commitment. The workability of the committee depended largely, over the twenty-odd years of its existence, upon the degree of amiable access of the science adviser to the Oval Office, a relationship that varied from one administration to the next depending, partly at least, on the personalities of the adviser and, of course, the president. The mechanism, useful as it was, came apart because of public disagreements by several committee members with what were at the time administration policies. Very likely it would have been killed off anyway, on grounds that the committee as a whole was thought by the White House to be a collection of elitist liberals embedded like a xenograft in the very flesh of the conservative government of the early seventies.

Nevertheless, PSAC was a good idea, and there are now signs that something like it will be put back in place under the chairmanship of Allan Bromley, the science adviser, whose direct access to the president seems now to be assured.

I raise the subject here because I believe that it is of more than provincial interest. Virtually all governments in the developed nations, and some of those still undergoing development, have in place one device or another for securing scientific advice to the executive, some of them informal and ad hoc, others highly organized and built into the structure of government. I suggest that the time has come for all of us to be thinking about how to put together an international body to serve the same function on a global scale, a planetary equivalent of PSAC, an international Science Advisory Council. I have no practical suggestions as to how to go about putting such a mechanism together, or how to select its necessarily representative and rotating membership. In a better world, such a body would be incorporated within the UN; I still think this would be a good idea if the committee or commission were to become a central part of the UN's responsibilities. It would provide the UN with one assuredly useful thing to do, very likely strengthening the role and position of WHO, another assuredly useful mechanism.

But the UN, at least for the time being, seems to be out of favor, its feet entangled in its own bureaucracy, and perhaps a new, totally independent nongovernmental organization, an NGO tied to no special region or interest, would do better what needs to be done. The International Congress of Scientific Unions should be considered for the assignment, or perhaps a small committee of ICSU members; the difficulty here would doubtless be the perception by many statesmen that such a group would be driven by the self-interest for which scientists have become so famous in recent years. I think I would vote for the UN, but with all kinds of reservations and misgivings.

Another possibility, thought to be unrealistic by the American scientific community until recently, is the International Institute for Applied Systems Analysis (IIASA). This highly intelligent and effective institution, with an admirable track record for the study of international science problems, was severely neglected by the U.S. government during the Reagan years, with the result that American participation became largely dependent on the unpredictable funds provided by various private, nongovernment organizations. Apparently, things have changed. The U.S. government is now providing funding for IIASA. The Bush administration is especially interested in IIASA's future research on global climate change. But it is a particularly encouraging item that the formal decision to fund IIASA does not specify the problems to be approached but represents a regular, predictable contribution to the core research problems to which IIASA itself is committed.

But something new surely needs to be invented and put in place. Perhaps some sort of international body modeled after the Science Policy Support Group, but charged with formulating an *international* rather than national agenda for science policy research, would be useful at the outset. Not only does the world of governments face huge and frightening problems needing the science of one discipline or another for solving—global warming or freezing, the ozone window, the population expansion, famine in more places than ever before, AIDS, malaria, trypanosomiasis, small wars, large wars, nuclear wars (perhaps the worst of all, *small* nuclear wars), one horrifyingly possible disaster after another, all in need of scientific inquiry, some perhaps only to be solved by new technologies not yet imagined. There is also another extremely serious problem: science itself could be going out of favor in the public mind.

There has never been a time of so much *magic*, all over the place, not just in southern California, and it seems to be spreading everywhere like a sort of infection. Intelligent, educated people sit cross-legged on cushions, surrounded by crystals, chanting, intoning, thinking deeply about deep breathing. They go to gifted readers to learn the future, or like drowsy presidents they read their horoscopes. People *channel*, never quite explaining what is meant by channeling. Everyone you pass on the street is on the way to visit a counselor, or to obtain something ineffable called therapy. They intone. The crystals have to be arranged in accordance with some special geometry, or the magic rays won't work. And all of this in aid of what it has become fashionable to call Wellness, more than just the absence of illness. Science, it is said, doesn't do any good because it is reductionist; holism is the way to go.

A couple of years ago some people in Lake Tahoe, on the California border, felt tired, and the new thing moved south and then east and now we have it in New York, the Chronic Fatigue Syndrome, CFS, believed to be caused by a virus, maybe EB. We have Post Traumatic Syndrome, PTS; Premenstrual Syndrome, PMS; Temporal-Mandibular joint syndrome, known everywhere as TMS; and, of course, in every city, town, and hamlet we have *stress*, deadly stress, deadly cancer-causing stress, sometimes caused by a fear of deadly cancer-causing dioxin (despite the fact that dioxin is mainly deadly for guinea pigs, as is, by the way, penicillin). I don't know what will happen when the word gets around that there are such things as quasi-crystals, mixed alloys slowly cooled, with stranger geometries resembling Penrose tiling, which means that the relationship between the not-quite-symmetrical parts are determined by the ancient golden mean, 1.618034, the basis of harmony in the Parthenon, the spiral fronds on sunflowers, the sequences in the Fibonacci series. More complicated crystal sittings, I suspect.

Doctors used to be the magicians, raising all sorts of expectations for health, but we are now magicians *manqué*, no good at crystal therapy, and untrained in the new discipline, not good enough, not the real thing. We are ineligible for holism, contaminated by reductionism, excluded from the world of the new science, needing replacement by "alternative" medicine.

And not just medicine. I sense, as a professional of sorts, a new atmosphere of anti-science, more than a fear of science, an anxiety to replace science by magic. I sense, as well, a general and sweeping anti-scientism, perhaps linked to anti-intellectualism as a new worldview, sweeping through the most educated and well-informed segments of the population. And, in my darker moments, I cannot think of anything to do about it except to wait in hope for it to pass away. Right now, however, we might as well recognize that anti-science is reaching the status of a philosophical position in the public mind, and we had better face up to it. I leave it there.

I take comfort from a paragraph at the end of a book by Peter and Jean Medawar, *The Life Sciences*. They write as follows:

> At no time since the early years of the seventeenth century have human thoughts been so darkened by an expectation of doom. In their apocalyptic moods people nowadays foresee a time when pressure of population will become insupportable, when greed and self-interest have despoiled the environment, when international rivalry has brought commerce and communications to a standstill.
>
> We, on the contrary, do not believe that any evil which may befall mankind cannot be prevented, or that any evil which now besets it is irremediable. For remedies people look first to science and then look away in disappointment, partly because they mistake the nature of the problems and partly because they have grown so used to thinking of science and technology as a secular substitute for the miraculous.

I agree with the Medawar appraisal, and would only add that there is no acceptable course for science to take for its future except, come what may, to keep at it.

We can be looking forward, if we are lucky, to bits of information about the world that will, as has happened from time to time in the past, change the way the world looks—even, if we become exceedingly lucky, change it for the better. Not in any everyday way, not in the growth of gross national products or the spreading around of prosperity, but simply in the way the world looks.

Something of this sort might be happening now, still at its beginning. There are good reasons to begin thinking about the earth as a kind of living organism. Despite the inner misgivings stirred inside some of our minds about the metaphoric word "Gaia," the notions proposed by James Lovelock for thinking this way about the earth are based on an array of solid scientific measurements.

I believe this new thing, that the earth is indeed a living organism, of greater size but probably no more complexity than any other attested biological organism, including our own human selves. We, for our part, are the equivalent of cells within the body of that creature. The whole thing breathes, metabolizes, adjusts its working parts (including, if we are lucky, us) automatically, autonomically, to changes within its internal environment. In any case, the surest, unmistakable evidence of coherent life, all of a piece, is its astonishing skill in maintaining the stability and equilibrium of the constituents of its atmosphere, most spectacularly and improbably the fixed levels of oxygen and, *pace* us, CO_2, the pH and salinity of its oceans, the diversity and developmental novelty of its kingdoms of live components, the vast wiring diagram that maintains the interconnectedness and interdependence of all its numberless parts, and the ultimate product of the life: more and more information.

One thing eludes me, always has and likely always will: If the earth is what I think it is, an immense being, intact and coherent, does it have a mind? If it does, what is it thinking? We like to tell each other these days, in our hubris, that we are the thinking part, the earth's awareness of itself; without us and our marvelous brains, even the universe would not exist—we form it and all the particles of its structure, and without us on the scene the whole affair would pop off in the old random disorder. I believe only a little of this, to the extent that whatever awareness we manage to achieve comes automatically into the earth's reach. But I believe another thing, somewhat larger. The earth consists of a multitude, a near-infinity of living species, all engaged in some kind of thought. Moths, for instance, do their kind of thinking; they have receptors for the ultrasound probes of bats, and swerve to one side or the other or drop to the ground if the bat is at a distance safe enough for that maneuver, but if the bat is close by, a meter or so away, and escape is nearly impossible, the moth does some very hard, quick thinking and switches chaos on in its brain. The result is a series of wild, unpredictable darting movements, and because of these an occasional, lucky moth escapes, and the bat is left with the second thoughts of bats.

Given brains all over the place, all engaged in thought, some of these to be sure very small thoughts, but all interacting and interconnected at least in the sense that the separate termites in a twenty-foot-tall termite hill are interconnected, and given the living mass of the earth and its atmosphere, including the swamps and the waters under the earth, there must be something like a mind at work, adrift somewhere around or over or within the mass.

My scientist friends will not be liking this notion, although I should think they would object to the less grand view that *any* system of living agents in close connection and communication with each other, sooner or later, when the mass of agents becomes large enough and dense enough, might begin to emit signals indicating coherence and moments of synchrony. Even so, my friends will object to the word "mind," worrying that I am proposing something mystical, a governor of the earth's affairs, a Presence, something *in charge*, issuing orders to this part or that, running the place.

Not a bit of it, or maybe only a little bit; my fantasy is of a different nature. It is merely there, an immense collective thought, spread everywhere, unconcerned with the details. It is, if it exists, the *result* of the earth's life, not at all the cause. What does it do, this mind of my imagining, if it does not operate the machine? It contemplates, that's what it does, is my answer.

No big deal, I tell my scientist friends; not to worry. It hasn't noticed you yet in any case. And anyway, if It has a preoccupation with any part of Itself in particular, this would likely be, as Haldane once remarked, all the various and multitudinous beetles.